A Revolution
Down on the Farm

A REVOLUTION DOWN ON THE FARM

The Transformation of
American Agriculture
since 1929

PAUL K. CONKIN

THE UNIVERSITY PRESS OF KENTUCKY

Copyright © 2008 by Paul K. Conkin
Paperback edition 2009

Published by the University Press of Kentucky
Scholarly publisher for the Commonwealth,
serving Bellarmine University, Berea College, Centre College of Kentucky, Eastern
Kentucky University, The Filson Historical Society, Georgetown College, Kentucky
Historical Society, Kentucky State University, Morehead State University, Murray
State University, Northern Kentucky University, Transylvania University, University
of Kentucky, University of Louisville, and Western Kentucky University.
All rights reserved.

Editorial and Sales Offices: The University Press of Kentucky
663 South Limestone Street, Lexington, Kentucky 40508-4008
www.kentuckypress.com

13 12 11 10 09   5 4 3 2 1

The Library of Congress has cataloged the hardcover edition as follows:

Conkin, Paul Keith.
  A revolution down on the farm : the transformation of American agriculture
since 1929 / by Paul K. Conkin.
     p. cm.
  Includes bibliographical references and index.
  ISBN 978-0-8131-2519-0 (hardcover : alk. paper)
  1. Agriculture—United States—History. 2. Agriculture and state—United
States—History. 3. Agricultural productivity—United States—History.
4. Agricultural innovations—United States—History. I. Title. II. Title:
Transformation of American agriculture since 1929.
  S441.C725 2008
  630.973'0904—dc22
  ISBN 978-0-8131-9242-0 (pbk. : alk. paper)

This book is printed on acid-free recycled paper meeting the requirements of the
American National Standard for Permanence in Paper for Printed Library Materials.

♾ ⊛

Manufactured in the United States of America.

Member of the Association of
American University Presses

Contents

Illustrations

Figures

Tables

Preface

When this book reaches an audience, I will be almost eighty years old. I was born in late October 1929, at the time of the stock market crash on Wall Street. My parents were scarcely aware of these events or of their significance. What they remembered was the first major frost of the year, which whitened the farm fields around the small, three-room cabin where I was born. My first opaque memory is of being led, probably by my father, down into the meadow, where I was greeted by neighbors who were helping to cut our tobacco. Because of the location of the crop (we rotated it every year), I later determined that this was September 1932, just before my third birthday.

It is appropriate that tobacco made such an early imprint on my memory, for it was our chief crop. Because it was a crop that required various tasks throughout the year, the cycles of work on tobacco were an important marker of the passing of time: gathering and then burning brush to sterilize beds for the plants; transplanting, cultivating, worming, topping, and suckering; cutting and hanging in the barn; stripping and grading the leaves; and finally the payoff—auction sales in December.

Until I was seventeen and in my final year of high school, I had assumed that I would make my career in agriculture. Four years of classes in vocational agriculture had idealized farming. But by my graduation after World War II, agriculture had begun a rapid transformation. It was clear that, at best, only a few local farmers could make a living on the rather hilly farms in our part of east Tennessee. Two of my teachers persuaded me to attend college (I was the only one to do so in my small graduating class of seventeen). Even though I spent my college summers working in the tobacco control program in my county, I knew by then that my career would be in education, not agriculture. Our small family farm survives. I now own it. I rent it to a neighbor for pasture and hay. It has been at least twenty years since anyone has grown cultivated crops, and even then it was less than an acre of land rented out for tobacco.

In what may be, because of age or health, my last book, I decided to go back to my beginnings, to the inescapable landscapes of youth, to go home again. My goal is to describe and explain the changes that have taken place in agriculture during my lifetime, to chart the distance traveled from what I experienced firsthand as a boy. Also, I want to clarify a complex story—the evolution of federal farm programs in that eighty-year span, for such policies provided the context for a revolution in agricultural productivity. It is also at the policy level that all American citizens are involved with agriculture, for it is their representatives in Congress who develop such policies.

Recently I challenged a former student who referred to the decline of agriculture in America. What could she mean by such a statement? It implies that something has gone wrong in the most critical sector of any economy. Certainly, from many perspectives, much has gone wrong. But as I hope to demonstrate in this book, agriculture has been the most successful sector in the recent economic history of the United States. The greatest industrial revolution in our history has occurred, with all its economic benefits and human costs, down on the farm, where productivity per full-time worker has increased at least tenfold since 1950.

This is a book based on memory as well as research. On one hand, I could not achieve my purposes without my own personal memories of growing up on a farm and my continued, if distant, involvement with what happened on that family farm. Throughout this book I rely on my past experience of farming and on my knowledge of the types of farming conducted by neighbors in our small community. On the other hand, I could not make sense of all those memories and experiences, place them in a broader context, or grasp their significance without having access to an enormous range of data, largely collected by the U.S. Department of Agriculture, and its subsequent analysis by economists and historians.

Agriculture, or at least the aspects of it devoted to the production of food, is the most basic of all human activities. Today, those who till the soil or tend livestock feed a world population of about 6.5 billion. Even fifty years ago the earth could not feed nearly so many people. And in fact, around 800 million hungry people are not well fed today. In another fifty years those who farm will have to feed an additional 3 billion. They will probably be able to do so, largely because of cumulative changes that occurred in my lifetime. What is sad, though, is that with each passing year, fewer and fewer local families can make a decent living on the farm. Today, my village is a far-out suburb of Kingsport, Greeneville, and Johnson City. Almost every adult has off-farm employment of some type.

In 1959 my sister married her high school sweetheart, whose family had moved from Tennessee to a dairy farm in York County, Pennsylvania. In this rich agricultural area, their farm managed to make all the needed adjustments to compete successfully. My brother-in-law remained a full-time farm operator until his death in 1994. During those years he bought a neighboring farm and increased his acreage to 400, making his farm one of the largest in the area. In later years even this was not enough land, so he rented more from his neighbors. He was able to upgrade his dairy farm to meet all the requirements of both the Philadelphia and New York marketing districts and managed to buy all the newest equipment, including several tractors and a self-propelled combine for both corn and small grains. In other words, he was a moderately successful commercial farmer. I thus use his experiences to illustrate some of the strategies that transformed American agriculture and reduced the number of American farm operators needed to produce 89 percent of our agricultural output from around 6 million in the 1930s to less than 350,000 today.

Acknowledgments

In writing this book, I drew on the knowledge of family members and friends, particularly my sister, Lois Conkin Hunt. Unlike in earlier books, I was not able to draw information and valuable criticism from students in my history courses, since I have retired from teaching. But I was honored to teach a brief, one-week course during the summer of 2007 in the Osher Lifelong Learning Institute at Vanderbilt. Most of the students were retired and old enough to have memories that stretch back to the Great Depression. They showed an intense interest in agriculture, offered helpful comments on my developing book, and asked questions that helped me better organize the chapters. I thank them. A friend with an extensive knowledge of agriculture, Thomas Bianconi, read the manuscript and offered several helpful comments. Above all, I want to acknowledge with gratitude an outside reader, the late Bruce L. Gardner of the University of Maryland, who read the developing manuscript twice and offered hundreds of useful criticisms.

Thanks to the Vanderbilt University Library, I have a wonderful library study. From the nearby library stacks or from our Interlibrary Loan service, I gained most of my research materials. No one could enjoy a more congenial work environment.

My illustrations reflect the generosity of several people. Pete Daniel, a curator in the National Museum of American History at the Smithsonian, a past student of mine, and a leading agricultural and southern historian, helped find four of the photographs from the museum's collection, including that of an original Hart-Parr tractor. Sherry Shaefer, who publishes the *Oliver Heritage* magazine, helped in the restoration of this Hart-Parr and made the excellent photograph included in this book. Harold Sohner created one of the replicas of the original McCormick reaper and generously provided the photograph for this book. This replica is in the Frontier Culture Museum in Staunton, Virginia. The excellent photographs of early Fordson and Farmall tractors are from the South

Dakota State Agriculture Heritage Museum in Brookings, South Dakota. Dawn Stephens, curator of photographs, helped locate the best possible photographs and provided them free of cost. The John Deere Company provided two photographs of very large John Deere combines.

As I wrote this book, I kept thinking of two pioneers in the field of agricultural history, a field that I have just invaded. We are all indebted to them for their lifelong work. One was a public historian, the late Wayne Rasmussen, who long headed the agricultural history section in the Department of Agriculture. The other, an academic historian, is Gilbert Fite, who knows more about farming than anyone I have ever known.

1. American Agriculture before 1930

From the beginning, English settlers along the Atlantic coast tried to find products that they could sell back in Britain. Many of these products involved the harvest of abundant forests or the purchase of furs from Indians. But very quickly the first colonists in Virginia found an exportable crop that was in great demand in Europe: tobacco. It was the first commodity, or money crop, for the new colony. It set a pattern. American agriculture from the beginning depended on markets. It was commercial.

Commercial Origins

Despite their commercial endeavors, most of what these early American farmers grew supplied local needs. Some refer to this as subsistence agriculture, but the label is misleading if it suggests that farmers, even in those first decades in America, supplied all their needs. They bought or traded for many items, including tools, housewares, exotic foods, and even some clothing and furniture. Native Americans had long exchanged agricultural goods for manufactured items, some procured from a considerable distance.

English colonists in North America simply adopted the same farming methods they knew from back home. They used the same draft animals and the same types of hoes and plows, and they planted the seeds they had brought with them to the New World. At first, they took advantage of the open land already cleared by the natives. Soon they added new land by clearing forests. They learned a few tricks from their Indian neighbors

1

and adopted Indian maize as the dominant cereal—more important than wheat, oats, barley, or rye in most regions of America. Unlike in Europe, land was plentiful, although it took hard work to get forestland ready for cultivation. Rents were low to nonexistent. It was labor that was expensive. Thus, American farmers sought laborsaving innovations, not the means to extract more production from each acre of land. In a sense, they were reckless in their clearing of trees and in risking both soil erosion and soil exhaustion. Like the Native Americans before them, they simply moved their crops to new ground when yields on older fields declined.

Unlike the Indians, the English settlers adopted a fee-simple type of tenure, not an open commons. Up through the nineteenth century, most American farms had more land than any one family could cultivate, even with many working children or one or two expensive hired hands. Only in the South, with its tobacco, coastal rice, and cotton crops, did Americans acquire a servile labor force that allowed the cultivation of large plantations—a pattern Europeans had earlier adopted in the Caribbean.

Through the early nineteenth century, Americans could best compete in world markets by selling farm and forest products as they grew or cut them or at the first level of processing (such as ginned cotton, flour, or lumber). Quite simply, they were able to produce such goods at the lowest possible prices. This involved not only a surplus of good land but also acquired skills and, gradually, the development of new and better tools. Americans exchanged such agricultural goods for more refined European manufactured products (furniture, fancy clothing, clocks) that, given the labor constraints, they could not produce at a competitive cost at home. This gradually changed after 1815, particularly in New England, as good land became scarce, labor supplies increased (young women were the first source, then immigrants), and new forms of manufacturing developed. Textiles led the way, as they still do in much of the world. They became the first major industry to adapt to a factory type of mass production.

In 1800 it took more than 50 percent of human labor worldwide to procure food. This was true even in England, despite its strong commitment to manufacturing. In the United States, one can only estimate the amount of labor devoted to agriculture. At least 90 percent of the population had some tie to agriculture, even if only part time. City artisans grew gardens and, if possible, owned cows, hogs, chickens, or all three. Lawyers, ministers, and schoolteachers almost uniformly owned farmland and at least supervised farming operations. At that time, given the level of agricultural technology, one farm family could supply food for only one other family on average. Also, farmers devoted much U.S. cropland to nonfood crops—mainly tobacco before 1800, but within two decades, even more so to cotton. But very slowly, with each passing decade, farming became more efficient.[1] By 1870, one able farmer growing wheat or corn could feed three other families. By 1900, one farmer might feed five other families; by 1930, almost ten. In 1900 agriculture was still, by far, the largest economic sector, with 40 percent of 76 million Americans living on farms. At least another 10 percent of workers were involved in agricultural services.

As a percentage of the total population, farmers declined year by year. The only possible exception was 1933 and 1934, when the normal pattern of migration to cities reversed, and the rural population increased. Meanwhile, the total number of farmers grew very slowly until the mid-1930s and then began a slow decline. In 1940 the farm population fell to what it had been in 1900—about 30 million (today it is around 4.5 million, but a majority of these people are only marginally involved in agriculture). The amount of cultivated land on farms increased until about 1935; since then, it has very slowly declined. But even in 1930 the largest segment of the labor force remained in agriculture—around 25 percent. By 1940, the agricultural revolution had begun in earnest, and the number of farmers and (less rapidly) the number of farms dropped precipitously with each decade until 1980. Both numbers have changed only marginally since then. However, the number of full-time farmers and the number of farms that contribute significantly to agricultural production (320,000 farms accounted for 89 percent of the total production in 2002) have continued to decline.[2]

The steady but slow growth of agricultural productivity before the Great Depression reflected several important innovations but could not conceal many deeply rooted maladies. The most important problem was the economic plight of the South in the post–Civil War years. Another was the growth of farm tenancy, which does not necessarily cause lower productivity but was in fact a major drag on production in the South. Another problem was the heritage of reckless exploitation of the land, with much of the South plagued by soil erosion and the overfarmed areas of the Great Plains subject to the terrible dust storms of the mid-1930s. In many places, overcropping had also depleted soil nutrients, a problem only partly solved by 1940 with the increasing use of chemical fertilizers. Another problem was price instability, with cycles of boom and bust leading to credit crises. Progress in farming increasingly depended on credit and thus the maintenance of debt as an operating cost.

Finally, as a whole, farmers were not very efficient. Almost everyone who writes about farm life stresses the hard work, even the drudgery. Farmers have helped nourish this image, often to their political benefit. But economists know better. Several farm tasks, particularly preparing the land for crops, planting, and harvesting, required long hours of work by several people, but only for short periods. Larger farms often had hired hands to meet these peak labor demands, but they had to board and pay these employees year-round. Clearly, farmers had to work more days out of the year than did city workers. Almost all farms before 1930 had horses to care for and feed twice a day. Most had milk cows, hogs, and chickens, meaning chores every day of the year. In this sense, the work was unrelenting, although it might take only a couple of hours each day. But aside from the peak periods, labor was usually redundant on farms. This meant that farmers often looked for outside work in the winter months, and southerners often rented out the labor of their slaves. Of course, farmers could find things to do on the farm in the winter or after the crops were laid by in the summer, but my observation is that they took plenty of time off for hunting and fishing or simply loafing at the country store. Also, apart from the intensity of labor during the harvest (or the threshing of wheat), I found the pace of farmwork to be leisurely, with rest periods, long lunch breaks, and the slow handling of more routine tasks. The amount of annual work performed by women easily exceeded that of men—child care, cooking, milking, churning butter, washing, ironing, sewing, tending a vegetable garden, and canning an endless array of foods for the winter. The cure for most of these problems lay ahead and would account for the takeoff in productivity after the mid-1930s—the main themes in subsequent chapters of this book.

In 1930, the average farm was not all that different from one in the late nineteenth century. Despite the increasing use of tractors, most farms still used horses and mules—about 25 million of them that year (down only 5 million from the peak in the early 1920s). By 1930, horse-based technology had reached its maximum development, and in some cases this was quite an intricate technology. Although farmers had increased their specialization, in the sense that most of them grew only one or two money crops, almost everyone had to grow corn (for human food as well as for livestock). Food for home use still came largely from the farm, as did the all-important fuel needed for cooking and heating. Gradual improvements in agricultural productivity before 1930 involved two factors: reduced labor for each acre cultivated, and improved yields per acre. The most dramatic improvements before 1930 were in labor requirements.

Tilling and Preparing The Soil

Historically, the most basic agricultural tool was the plow. Before plows, the ground had to be prepared by sticks, hoes, or spades. The predecessors of modern plows go back to ancient Egypt and Mesopotamia. The curved trunk of a small tree made up the beam, pulled by oxen, and either a sharp curve in the beam or a second strip of wood tied to the beam was sharpened at the end and used to make a furrow in the ground. Handles were also attached to this crude plow. Of course, the wood point quickly became worn and dull, so the Egyptians made a hard cover from flint. By biblical times, iron was used for what we now call the share. In some light soils, furrowing plows work fairly well, but not in tight soils or in sod. The modern equivalent of these first plows was what we called, in my youth, the single-foot plow, which we used to create rows in the garden and in tobacco fields. The Romans added larger points or shares, as well as a flat board along one side of the share to force the disturbed soil to move in one direction. This was the origin of what became the moldboard and a true turning plow, one that flipped the sod over in such a way as to loosen the disturbed dirt.[3]

By 1700, the English had devised dozens of plow designs, most of which involved wooden moldboards that forced some turning of the soil, but these were usually sloped or only slightly curved. They were hard to pull and left the sod vertical to the ground, not fully flipped. The first moldboard with the double curve of a modern plow was perfected in England in 1730, and this type of plow sold well. Almost all plows used in America before 1800 had wooden beams and moldboards and cast-iron shares, which dulled quickly. But after the American Revolution, new designs appeared almost every year. Thomas Jefferson invented a new moldboard (curved, but still too flat) and proposed to replace his wooden model with one made of cast iron. The first American patent for a cast-iron moldboard was issued in 1800. By 1814, American inventors filed patents for plows with cast-iron beams and moldboards. These had curved moldboards close to those of modern plows, and most had a colter attached to the beam, in front of the moldboard, to cut through the sod. By then, several inventors had found ways to bolt the plowshare—the cutting edge—to the moldboard, allowing farmers to replace a broken or worn share quickly, but the intractable problem was finding a moldboard and share that easily scoured (shed the dirt). In 1837 a blacksmith, John Deere, became famous when he replaced cast-iron shares with ones made of steel, which was still quite expensive. Steel shares provided the polished, dirt-shedding surface needed to turn the deep virgin soils of the Midwest. When steel became cheaper, he and others added steel moldboards and, after the Civil War, even steel beams.

In most soils, the single plow required two horses. On small farms, the two-horse plow remained dominant until 1930, or when horses began to give way to tractors. On larger farms, beginning as early as 1864, the standard plow became a riding plow with two wheels and a seat, often called a sulky (after harness racing). Four horses pulled it. At first it had only one moldboard, but inventors soon added a second. By World War I, an increasing number of farmers in the West, Great Plains, and Midwest did their plowing with heavy tractors that could pull four or more bottoms, which they called gang plows.

Except in the most friable soil, plowing did not leave the soil ready for the planting of seeds. Further preparation required leveling and breaking up clods—a process called harrowing. This was not a complex task. Some farmers used a simple tool called a drag—boards nailed to a frame and pulled over the soil, with stones or even children added to weigh it down. Later, heavy rollers did the same work. Even in Roman times, farmers used brush from trees to drag over the soil. The Romans also used wooden pegs attached to a wooden frame to pulverize soil, or the first true harrow. By 1800, farmers had developed dozens of different designs and shapes for harrows and had replaced wooden teeth with iron spikes. When I was a boy, the favorite local design was a double A-frame, with iron teeth. By then, more sophisticated harrows had spring teeth that could better hug the ground. However, no toothed harrow was as efficient as the disk harrow, which was perfected just after the Civil War. These implements had revolving concave disks that rolled through the soil at an angle determined by the operator (the greater the angle, the more the disks dug into the soil). Later, large disks pulled by tractors could cut so deeply into soil as to become substitutes for turning plows. Such disks are still used today, although they are not as effective as revolving tines (called rotary hoes or, for gardens, rototillers).

TOOLS FOR PLANTING AND CULTIVATING

Seed planting followed soil preparation. From ancient times, farmers did this by hand. They broadcast seeds of wheat and other small grains by hand and dropped the seeds for row crops into hills or prepared rows. The challenge was to find machines that did these jobs better and faster. For most of the grasses used for hay, the goal was a mechanical broadcaster. The earliest answer was a hand-held and hand-cranked hurricane spreader (I still own and use one of these). Later, farmers attached rotary spreaders to the backs of wagons, with power supplied by a wagon wheel. For the spreading of fertilizer and lime on pastures, large hurricane-type spreaders, operated by a tractor or on the back of a truck, are still the preferred tools. Broadcasting worked best for seeds that did not require a covering of soil to germinate. When used for small grain, one had to use a harrow or brush to cover at least most of the seeds, to protect them from birds and to ensure enough moisture for germination.

What was needed was a drill. It took more than two hundred years and hundreds of experimental designs to perfect a single-row drill or planter for row crops, and multiple drills, placed side by side at about eight-inch intervals, for small grains. In each case, the challenges included making a device to feed seeds from a hopper at the proper rate into a funnel or tube leading to the ground, a device to open a small furrow at the desired depth as a receptacle for the seeds, and a device to cover the seeds with soil. In addition, in a grain drill, with its width and multiple feeds (about eight in early drills; twenty or more in those pulled by tractors), the feet that work at the ground level have to be flexible enough to maintain contact and the correct depth over uneven ground. This requires either springs on each tube or a hydraulic device (like a shock absorber on an automobile) to compensate for uneven surfaces and to spring back from rock or roots. Drills that met some of these requirements were in use by 1840, but it was only after the Civil War that drills meeting all these requirements were available at a price farmers could afford. Later drills, like most contemporary ones, had double disks on each head to open space for the seeds. By 1900, drills usually had three bins—one for the grain, another for fertilizer, and one for grass seed that was often planted with wheat. By 1930, on moderate-sized, pretractor farms, the drill was usually the most complex farm implement except for the binder for small grains.

For row crops, the early drills or planters used a single wheel to provide the power. In time, almost all such planters would have revolving disks with holes that captured the seeds and dropped them when the hole passed over the funnel or top of the tunnel. Various hole sizes and various intervals between holes fit different crops and allowed different spacing. By the Civil War, most large farms used two-row planters for corn, with two wheels and a seat for the driver. In the Midwest, as part of a weed control strategy, the two plates were synchronized (check-row planters), so that the seeds (usually three or four) could drop at roughly three-foot intervals, creating a checkerboard pattern. This meant that a cultivator could plow in two directions, almost eliminating the need for hoeing.

For some crops (tobacco, sweet potatoes, tomatoes), the most labor-intensive task was the transplanting of seedlings. Either the seeds had to be started indoors early, before the end of spring frosts, or the tiny seeds did not lend themselves to field conditions. At the appropriate time, the tender plants had to be transplanted in the fields, which was backbreaking work. Only after World War II did transplanting machines, pulled by either horses or tractors, lessen the drudgery. The machine opened a furrow, and two "droppers" rode on the moving machine and alternately placed a plant in the row. The machine then both watered the plants from its large water tank and closed the dirt around them.

After planting, row crops had to be kept clear of weeds, which was one of the great challenges of agriculture. This had long been the main task for hoes, but the arduous work of hoeing corn, cotton, or tobacco cried out for a mechanical replacement. This meant horse-drawn plows that could cultivate between rows. It took a long time to perfect cultivating plows, but by 1820, farmers were using plows with two large feet (or shovels), and later plows with up to five smaller feet. As a boy, I plowed corn with a three-foot cultivator with shares or points of moderate width; differently shaped points were available for different tasks. To cultivate a row crop, one would run the cultivator as close as possible to the plants, removing almost all the weeds except those actually in the row. By the Civil War, farmers had riding cultivators (pulled by two horses) that straddled the rows. When they ran such cultivators in both directions in cross-checked corn, they almost eliminated the need for hand hoeing.

Tools of Harvest

For most crops, harvesting was the most labor-intensive chore and therefore the greatest challenge for inventors. For small grains, harvesting involved three tasks: cutting the stalks of grain with their golden ripe heads, separating the grains from the head (threshing), and winnowing out the chaff that remained after threshing. For a thousand years these three tasks involved the same labor-intensive methods. Farmers used a sickle or a scythe to cut handfuls or swaths of grain, leaving it to others to collect, bind, and shock the grain. When the grain was fully dry, they separated the seeds from the head by using flails or by having horses or cattle tramp them out. Finally, on windy days they poured the grain back and forth between containers until all the chaff blew away, leaving a reasonably clean product. Eventually, new tools reduced this human labor by a factor of fifty. It is one of the better-known stories in agricultural history.[4]

The first laborsaving device may have originated in America, although we do not know who perfected it. By the American Revolution, the cradle was slowly replacing the scythe, which had largely assumed its modern form. The scythe had a long, curved steel blade with a thickened strip at the back edge to give it strength. The handle had a double curve, which allowed the operator to stand at a normal height, holding on to two nibs or wooden handles along the snath. This was a marvel of engineering in itself. The cradle had a light wooden lath with six or more curved wooden fingers attached to the base of the snath. The fingers matched the length and curvature of the blade. As the cradler swung the device through a long sweep, the falling grain collected on the fingers, allowing the operator to dump it in a neat pile for the shockers, who came along to collect and bundle the grain. A trained cradler could cut an acre a day, and a few young men claimed that they could cut up to five or six acres. I can still remember my grandfather cradling a field of grain when I was a boy. The cradle remained the main tool for cutting grain until after the Civil War, but by 1835 a few large farmers had turned to a much more complex and costly cutting machine—the reaper and, with later refinements, the binder. These machines eventually transformed the harvest of small grains. Of all nineteenth-century agricultural inventions, only barbed wire was close to being as revolutionary, since it gradually displaced rail fencing and transformed the range cattle industry.

After years of experimentation by several inventors, Cyrus McCormick built one of the first almost successful grain reapers in 1831 (see figure 1). Obed Hussey had a working reaper by 1833 and obtained the first American patent. His machine had the best cutter bar. The basic concept was thus established by 1835, but workable machines—ones that did not break down and did not cost too much—evolved slowly. By 1860, they were at work on a minority of farms (60,000), but they were clearly the technology of the future. The secret of a successful reaper was the cutter bar, which revolutionized not only the wheat harvest but also the cutting of hay. It is still used on contemporary grain heads for combines. The reaper had only one engaged, lugged wheel directly under the weight of the machine (called the bull wheel), which transferred power by pulleys and chains to a revolving wheel with an attached lever that propelled the bar back and forth, much like the drivers on a steam locomotive. It was, and still is, a very effective cutting device, but it allowed very little toler-

Figure 1. An exact replica of Cyrus McCormick's 1831 reaper. (Courtesy of Harold Sohner)

ance in the various parts. One other innovation of the McCormick reaper (but not Hussey's) was a revolving reel above the bar. It had wooden slats and moved the grain into the cutter bar.

The reaper's early versions did not drastically reduce labor. The cut grain fell onto a stationary platform, and a worker had to use a hand rake to pull the grain off the end for the shockers who followed. An early innovation was an automatic raker. Standard after 1864 was a canvas on rollers that moved the fallen grain up to a platform on the reaper, where workers gathered it into bundles, tied together by twisted straws. By 1873, an automatic mechanism tied the clustered wheat with wire, but the wire was a deadly hazard in threshing and for livestock. At least two inventors patented knotting devices that would bind the grain with twine. In the early 1880s, both the Deering and the McCormick reapers used this device on the first successful binders. The binder remained the most advanced technology for the harvest of grain in much of the eastern United States until after World War II. Not so in the drier wheat areas of the West, particularly in California and eastern Washington and Oregon. There, a machine called a header dominated the harvest. It consisted of a wide cutter bar and reel, with a rolling canvas that moved the short-cut wheat into special wagons to be hauled to stationary threshing machines. The machine was called a header because horses pushed the grain head from behind. It worked only with very ripe, dry wheat and eliminated the binding and shocking of wheat.

After the invention of the reaper, inventors tried to find a way to move a threshing machine through the fields and cut and feed grain into them. Numerous patents for a combined implement ensued, and one 1839 model in Michigan cut and threshed wheat but did not clean (or winnow) the grain. This type of machine, to which cleaning blowers were soon added, was first adopted by wheat farmers in California before the Civil War. As late as 1880 these huge combines still harvested only 10 percent of the crop in California, but they quickly replaced headers and threshers in much of the West over the next two decades.

The combine, the most intricate and costly farm machine invented in the nineteenth century, used the same cutter bar and reel as the binder. It fed the cut grain into the large, efficient threshing machines that had been perfected by 1880. The early, heavy combine was practical only on the very large, dry, level wheat fields of the West. In 1891, in California, Benjamin Holt invented a leveling device for combines. His Holt Hillside combine made mechanical harvesting possible in the hilly Palouse region of eastern Washington. The aridity of the West meant short wheat and less falling over than occurred with heavy-headed, dead ripe wheat in the wetter eastern states. The early combines were pulled by from twenty to forty horses. These combines, and the horse hitches used to pull them, were marvels to behold (see figure 2). In hilly areas, horse-drawn combines continued in use until the 1930s. Some of these machines were the largest ever employed on American farms. One inventor in Stockton, California, created a version of a self-propelled combine in 1886. In the center of it was, in effect, a detachable steam tractor. Its first head measured twenty-two feet and, after 1888, forty feet, or as large as on any combine used today. After 1890, in level terrain, combines were increasingly moved by huge steam traction machines, some weighing more than 40,000 pounds. Only after 1940 were combines widely used in the eastern United States. This required the development of both lighter combines and electric or propane dryers that allowed wheat elevators to accept wheat with too much moisture, with the cost of the additional drying deducted from the selling price.

New threshing and winnowing tools made possible the modern combine. Threshing machines actually preceded the invention of the reaper.

Figure 2. Five Holt Brothers Hillside combines, each pulled by twenty-seven horses or mules, near Walla Walla, Washington, 1908. (Smithsonian, National Museum of American History)

After the American Revolution, inventors perfected dozens of threshing machines. Some used hand- or horse-powered flailing devices; others used revolving rollers with teeth or spurs. In time, rollers prevailed. Horses provided power for the larger machines by means of a treadmill or by walking in circles (as for old sorghum mills) to turn overhead pulleys attached by a belt to the thresher. These machines did not do a good job of cleaning the grain. For this, winnowing machines were required, and again, dozens of models were patented before the Civil War. They involved a combination of two technologies—shaking or vibrating screens, and fans. A next-door neighbor had such a nineteenth-century machine in his barn when I was a boy (he still has it). Wheat is poured in at the top; it then falls through a series of variously sized mesh screens while a fan blows through the device. By the time the wheat falls to the bottom, it is clean. My neighbor's is a hand-cranked machine, but horses pulled larger ones.

In 1844 J. I. Case marketed the first thresher that was able to winnow the grain. It became the standard model after the Civil War, and Case remained the largest producer. Initially, these heavy threshing machines were powered by up to eight horses. Some of these horse-operated threshers remained in use into the twentieth century. First, a set of rollers separated the grain from the straw. Then screens and a powerful fan winnowed it. The clean grain fed into an outlet to be bagged or gathered in a wagon or truck. By 1870, a few large steam engines, mounted on wheels, began to power threshing machines, and after 1880 they dominated threshing. At first, horses had to pull these land locomotives, but by 1890, most were self-propelled, either by a chain from the belt pulley to one lugged wheel or by a gear system with only one speed—very slow. They could pull themselves and a heavy thresher from farm to farm, but they could not travel on paved roads, where the cleats could do serious damage, or over most bridges. Early versions expelled the straw at one end, where rakers removed it with pitchforks. Later machines had a net or canvas device that carried the straw out and piled it up. By 1900, the fans on most threshers were powerful enough to blow the straw and chaff out a long, maneuverable wind stacker, allowing the operator to create a beautiful cone of straw—the straw stack of legend (see figure 3). As a boy I looked forward to threshing time, for it provided us with fresh, buoyant straw that we used to refill our straw ticks, at a time when poor families could not afford mattresses.

Figure 3. A restored Rumely Oil Pull steam tractor powering a threshing machine. (National Museum of American History)

Individual farmers could not afford this type of threshing machine. They simply could not use it often enough to justify the high cost. Thus, threshing in the late nineteenth century became the first major customized task in American agriculture. Custom threshers bought their own machines and moved them from farm to farm, often in a pattern repeated year after year. They received money or a portion of the threshed grain in payment. In some areas, multiple farmers formed a threshing network and collectively owned a threshing machine. Threshing required at least five workers to operate the machines: one to cut the twine on the bundles, one to feed the bundles into the machine, one to operate the stacker, one to oil and care for the machines, and one to bag the grain at the end of the process. It took at least this many, and often more, to keep the wheat coming to the machine. Thus, on threshing day a crew of at least ten followed the thresher, assured of work at a higher than normal wage and a fabulous meal supplied by the farmer's wife. These meals became justifiably famous, a point of honor and a type of competition among the wives. I followed the thresher for at least two years as a teenager. The work could be backbreaking, but I often received as much as $3 a day (at the time, standard pay for normal farmwork was $1).

Unlike the small grains, the harvesting of most row crops was largely unaffected by the invention of new tools before 1930. Only corn benefited, but even there the changes were just beginning. Both cotton and tobacco required an enormous amount of hand labor to harvest. Tobacco still does. And as late as 1930, no machines had replaced the human hand in picking cotton bolls. The revolutionary cotton picker was not widely used until after 1950.

Corn was never as labor-intensive as either cotton or tobacco. Two modes of harvest were typical by 1920. When farmers wanted to preserve the dried cornstalks for fodder or to clear the land for the planting of winter wheat, they cut the stalks by hand and clustered them in shocks, like the ones seen in Currier and Ives prints. The other harvesting technique, which dominated in much of the midwestern Corn Belt, involved shucking the corn in the field. Horses pulled a wagon alongside a row or two of corn, and pickers used gloves and a small steel hand tool (called a peg) to open up the shuck and get a clean ear, which they tossed into the wagon. Well-trained horses would keep the wagon in the correct position without a driver. The corn was hauled to a slotted crib (the slots allowed circulating air to dry the corn), and the stalks were left for cows or hogs to feed on in the field. This method required less labor than shocking. My father went to Iowa in the fall of 1920 for the shucking (also called husking), which paid well. He did not observe any corn picking machines, although a few were in use by then. As late as 1930, more than 90 percent of corn was still shucked by hand.

The Tractor

Among the many innovations in farming, the only one that would replace horses was the tractor. Its enormous impact came largely after 1930, but by then the possibilities were already clear. For many farmers in 1930, a costly tractor might have saved some labor, but it would not necessarily be cost-effective unless it could replace—not just supplement—horses and mules. Within the next decade, it would be able to do so on most farms; within the next two decades, it would do so on all farms except those in mountain valleys or those farmed by the Amish.[5]

The word *tractor* derives from the term *traction machine*, used for early self-propelled steam engines. A few experimental and at least partly self-propelled steam engines were in existence just after the Civil War, but they were not widely used until after 1880. At that time, they were employed almost exclusively for pulling threshing machines, for which they provided the necessary belt power. In the 1890s on the Great Plains, such self-propelled traction machines pulled both plows and early combines. They were much too heavy for other farming tasks, since they severely compacted the soil. Meanwhile, one-cylinder internal combustion engines were available after 1870 and were useful for some farm tasks, such as milling grain or sawing wood. The first known use of such engines for traction work came in 1892, when an internal combustion engine was substituted for steam to operate and move a threshing machine. The manufacturer (a predecessor of the John Deere Company) made no marketable machine, concentrating instead on producing the new type of engine. But several inventors accumulated different parts and made what was, by then, called a tractor. The Hart-Parr company (later absorbed by Oliver) placed the first production tractor on the market in 1902, with fifteen sold in 1903. A restored version of this early Hart-Parr tractor is now on exhibit in the Smithsonian (see figure 4). In form, it looks like a contemporary tractor, but this was not true of any number of experimental designs created over the next twenty years. In 1906, among dozens of competitors, the new International Harvester Company (a merger of the McCormick and Deering farm machinery companies) sold its first tractor, and it would build a thousand a year by 1910. This early boom ended by 1912, when the

Figure 4. Hart-Parr No. 3, the first commercial tractor. (National Museum of American History, photograph courtesy of Sherry Schaefer)

very small market was saturated. These early tractors were often a disappointment to their owners.

About the only profitable use of these early, very expensive tractors was for plowing, threshing, or pulling combines. They were huge machines, often weighing more than 20,000 pounds (one Hart-Parr tractor weighted 35,000 pounds). Some were powerful and could pull four or more deep plows through virgin prairie soil. But breaking the plains was a one-time operation, and they were too heavy for use on already cultivated land. Even as late as 1917, there were fewer than 80,000 tractors on American farms. But by then a series of innovations, as well as some lighter tractors, had created an expansive market, leading to a brief boom during World War I. By 1918, tractors were switching from kerosene to gasoline, a few already had electric starters and lights, and a Moline model had a custom-made plow (most tractors had simply pulled existing horse-based implements) with a power lift. By 1921, American farmers owned more than 300,000 tractors.

The largest impact on the market was Henry Ford's introduction of the mass-produced Fordson tractor in 1917 (see figure 5). It was not

Figure 5. 1918 Fordson, the first mass-produced tractor. (South Dakota State Agriculture Heritage Museum Photograph Archive)

a marvel of engineering. The Fordson had a dangerous worm drive (it easily reared up when it hit obstacles and in this way killed many operators) and a cheap magneto (it was hard to crank), but it was smaller (it pulled only two plows) than most tractors on the market, and it was comparatively inexpensive, at an initial wartime price of $750. Ford had sold more than 100,000 tractors by 1920. In 1922, during a price war with International Harvester, Ford lowered his price to $395—or what he claimed was the cost of a good team of horses. He lost so much money that he stopped making tractors in America in 1928. My father, who worked for two years on a farm in Iowa in the early 1920s, encountered his first Fordson there. The owner never let my father, or even his own sons, drive his new pet, which he used only for plowing. I doubt that it was cost-effective, in large part because, with such a restricted use, it could not replace any of the farm's four horses. But by 1925, at least in the Wheat Belt, tractors were profitable investments simply because the newer models could do all the soil preparation, drilling, and combining, or all the work required for small grains.

What eastern farmers needed was a small, inexpensive, all-purpose

tractor that could cultivate as well as plow, was easily maneuverable in
small fields, could directly power fully integrated tools, and could pull
wagons. Such a tractor could replace horses and mules, thus freeing up
almost 25 percent of all land under cultivation. A series of innovations in
the 1920s almost met these requirements. The power takeoff, developed
in France before World War I, was first sold on an American tractor in
1922. It eliminated the bull wheel on binders and mowing machines,
and when the shaft and speed became standard, it provided the main
source of power for almost all farm implements. By then, improved air
filters had eliminated an early problem with dust-clogged engines. In
1924, International Harvester finally introduced a tractor that could cultivate. It had the beguiling name of Farmall (implying that it could do all
the tasks on a farm). It was narrow, was high enough off the ground to
straddle growing row crops, and had both front and side attachments for
cultivators. Its most distinctive feature was a pair of closely joined front
wheels that could fit between rows (see figure 6). In a competitive model
introduced in 1927, John Deere added a hydraulic power lift to allow
easy control over new equipment designed for tractors. By 1930, at least
three companies—including the very successful Allis-Chalmers—had introduced even smaller one-plow tractors. One, the Farmall A, is still in use

Figure 6. 1928 Farmall with attached two-row cultivator. (South Dakota State Agriculture Heritage Museum Photograph Archive)

on some farms. Thus, in most respects, the modern all-purpose tractor had reached its maturity by 1930, with 1.2 million in use (6.8 million farms still had no tractors; many could not afford one). But clearly, one reason for the dramatic rise in farm productivity in the 1930s was the increased use of tractors (with rubber tires after 1935 and a three-point hitch after 1940) and the gradual replacement of horses and mules.

RESEARCH, EDUCATION, AND EXTENSION

Without supportive governmental policies, the success of pre-1930 American agriculture would have been impossible. These policies included generous terms for the disposal of public lands; federal and state funding of roads, canals, and railroads; and land grants for public school systems and colleges. After the Civil War, this support increased and soon included government-supported research, regulatory legislation, subsidized irrigation projects, and accessible, low-interest credit. In the 1920s, major debates about farm policies led to several new initiatives and helped prepare the way for the farm legislation of the New Deal, which, for the first time, directly supported agricultural prices well above free-market levels.

As early as 1862, with the Morrill Act, the federal government assumed primary responsibility for agricultural research and development. Given the large number of individual proprietors in agriculture, only a public agency was in a position to undertake such research. The Morrill Act gave each state 30,000 acres of public land for each of its senators and representatives, with the proceeds from the sale of this land to become an endowment for an agricultural and mechanical college. Three or four agricultural colleges already existed, but they needed additional funding. When these original endowments proved insufficient, Congress added annual appropriations that continue to the present. Today, these land-grant institutions include more than half of the largest and most prestigious public universities in the United States. Although the federal funds supported instruction in such fields as home economics, industrial arts, and engineering, farmers were the major beneficiaries. Today, the colleges of agriculture in these land-grant universities constitute the largest pool of agriculturally related scientists in the world. The second largest pool is probably in the U.S. Department of Agriculture, also founded in 1862.

The original Morrill Act did not make any provision for African American farmers in the South. All the southern land-grant colleges were segregated. This concerned its sponsor, Justin Smith Morrill of Vermont, and other northern legislators. Thus, in 1890 Congress amended the first Morrill Act by adding funds and requiring an end to discrimination against blacks. Within a few years, all the former slave states established separate black land-grant colleges (often referred to as the 1890 colleges). A few states granted at least a part of these funds to private colleges, with Alabama distributing most of its funds to Booker T. Washington's Tuskegee Institute. These new agricultural and mechanical colleges became in time the first public black colleges and universities in the South. But in the first fifty years, few of these struggling colleges were able to hire trained agricultural experts or to attract many students to what most blacks viewed as stigmatized vocational education. Most of these new colleges became, in effect, normal schools for blacks, with the level of instruction rarely above a high school level. The two exceptions to this sad story were Tuskegee and Hampton Institute in Virginia. Both soon had able staffs in agricultural and industrial areas and would train almost all the early black scientists in these fields.

Soon after the first Morrill Act, several new agricultural colleges began rather extensive research and demonstration efforts in what were called experiment stations. In 1883 the Department of Agriculture joined in this early research, with an experimental farm on the Mall. This later moved to Arlington, Virginia, and in 1910 to the present home of the National Agricultural Research Center in Beltsville, Maryland. In the Hatch Act of 1887, Congress authorized funds to support an agricultural experiment station in each state. This act preceded the 1890 Morrill Act revisions and made no provision for black experiment stations (the first federal funds for agricultural research in the black land-grant institutions came in 1967). According to the Hatch Act, these stations were to carry out original research, investigations, and experiments comparable to those in the manufacturing sector, with the objective of supporting an effective agricultural industry in the United States and improving rural life. To further these goals, the stations were to publish and distribute (with free postage) to farmers their bulletins, reports, and articles, as well as provide on-site demonstrations of the latest agricultural methods. The original appropriation was modest—$30,000 to each state, paid for by public land sales. Later, direct appropriations would take over as land sales declined. The scientific staffs of such stations were members of the faculty of the parent college.

An Office of Experiment Stations in the Department of Agriculture set the rules for the use of federal grants. At first, scientists on the faculties of the universities undertook the research task, while still teaching their regular courses. Each land-grant university either used part of its existing campus or bought new land to establish an experimental farm. The federal office pushed the stations to engage in basic research, whereas farmers sought direct services, such as the testing of competing fertilizers. Farmers also requested the establishment of branch stations to improve their access to the demonstration work. Until after 1900, the Office of Experiment Stations opposed such branches, but several states used state funds to establish them. Soon, almost all state experiment stations had regional branches. In 1906 Congress passed the Adams Act, which doubled the federal funding, but with the stipulation that at least half the funds be expended for original research. In the 1920s the federal funding for these stations tripled, with some states receiving more than $1 million each year. From the beginning, the stations carried out work on soils, insects and diseases, new crop varieties, livestock improvements, and land use. The stations also tested soil for local farmers, offered advice to housewives on home management, helped farmers manage their forest resources, and, perhaps most important of all, used their own fields to demonstrate the latest farming methods.[6]

By subsidizing research and development in the experiment stations, Congress tried to ensure the rapid dissemination of the most advanced farming practices. For the largest farms and the best-educated farmers, this tactic undoubtedly worked as planned. But these stations had little effect on average farmers, most of whom lived a considerable distance from a college or even a branch station and did not read the literature emanating from them. Thus, other means of extending agricultural knowledge to farmers were needed. Early efforts included correspondence courses offered by land-grant universities and demonstrations and courses offered by railroads. But the major tool was farmers' institutes, modeled on the large number of institutes for public schoolteachers. By 1900, most rural counties were organizing weekend or weeklong institutes for farmers, who came to hear lectures, attend demonstrations, and learn the latest farming techniques.[7]

The one person most responsible for organized extension work was Seaman A. Knapp. Born in New York, Knapp was college educated and had an early career in teaching. He moved to Iowa in 1865 for health reasons and became a hog farmer, clergyman, school superintendent, and editor of a farm journal. He joined the faculty of Iowa Agricultural College in 1879 and briefly served as its president. There he helped gain support for what would become the Hatch Act. In 1885 he moved to Louisiana to help a land syndicate develop a new inland Rice Belt. This new capital-intensive crop led the way toward a scientifically informed agriculture in the South. To facilitate rice cultivation, Knapp helped set up demonstration farms and encouraged local farmers to visit and learn from them. In 1898 the Department of Agriculture appointed him a special agent to travel to the Far East to seek better varieties of rice. After two trips, he came home and accepted an appointment in 1902—the first of its kind—as a special agent to promote agriculture in the lagging South.

In a sense, agricultural extension began with Knapp's work in 1903. He instituted a new type of demonstration farm in Texas, recruiting an ordinary farmer to lend part of his land for the demonstration of new varieties, fertilizers, and methods. Community leaders agreed to fund any losses occasioned by the experiments (however, the farmer actually profited from them). This was the model for what eventually became hundreds of such farms. Also in 1903, Congress appropriated funds for boll weevil control and gave some of it to Knapp. He used his demonstration farms to test new methods of control and in 1904 used his funds to employ temporary agents to spread this knowledge directly to farmers. By the next year, he had twenty special agents working in Louisiana and Texas. These agents were farmers, not college-trained experts; they worked for two-month terms and received $60 a month.

Knapp soon expanded the work of his agents from boll weevil control to all aspects of farming. Although he employed a professor from Texas A&M as an agent in 1909, he usually kept his extension service separate from the agricultural colleges and experiment stations. With some re-luctance, in 1906 he appointed the first African American agents—one from Tuskegee and a second from Hampton Institute. In 1906 the new Rockefeller-funded General Education Board began funding Knapp's work, with the annual amount growing to $187,000 by 1914; that same year, federal appropriations reached $335,000. By 1913, two years after Knapp's death, the system of extension agents had reached all the southern states, while some northern states funded imitations. The number of agents grew from 49 in 1907 to more than 700 in 1912, 32 of whom were black. By then, most agents served in only one county and were usually referred to as county agents. Most received supplemental funds from states, local governments, or business organizations. By 1912, an increasing number of states had given administrative control of the program to the land-grant universities. Also, many county agents had established boys' clubs, often tied to projects in a specific crop (for instance, many were called corn clubs and required a boy to grow one acre of

corn); a few had also formed girls' clubs. These were the predecessors of
the 4-H Clubs formed by the new Cooperative Extension Service created
in 1914 by the Smith-Lever Act.[8]

The Extension Service was, in effect, the final expansion of the Morrill
Act. In addition to operating experiment stations, the land-grant agricultural colleges now had a legislative mandate to carry new knowledge in
agriculture and home economics directly to farmers and homemakers
in each rural county. As implemented, the Extension Service became the
most ubiquitous facet of the huge Department of Agriculture bureaucracy. In all counties with a rural population (most in 1914), the service employed extension agents to work directly with local farmers and
homemakers. These agents were almost always graduates of a land-grant
university and had faculty status in such a university; they had offices
in the county seat but were often at work out in the field in rural communities. Agents helped set up home demonstration clubs in most farm
communities (my mother faithfully attended monthly club meetings for
two decades). They sponsored local demonstrations for farmers; gave free
advice about fertilizers, seed varieties, and livestock breeding; and eventually aided farmers in the sale of timber. The same southern states that
had refused to establish black experiment stations followed Knapp's precedent and hired black extension agents. After some bitter debates, southern congressmen who had successfully resisted amendments to the act
that required black access promised to set up a separate Negro Extension
Service. They did so, but it was the white land-grant college that selected
the agents. Typically, African American agents served a larger constituency than white agents and received only half their salary.[9]

A final strategy—essentially one of adult education—helped alle-
viate the limited effectiveness of extension programs over succeeding
generations. This was the Smith-Hughes Act or, formally, the National
Vocational Education Act of 1917. It provided federal subsidies for vocational education in secondary schools, which were finally becoming
well established in most states. Except for the early land grants, this was
the first major federal aid for public education in America, and it would
remain the only such aid for the next forty-five years. The act established
two funds to help pay teachers and their supervisors in vocational education. One fund, for agriculture, amounted to $500,000 in 1918, increas-
ing incrementally to $3 million in 1926 and for each year beyond that.
This fund was apportioned according to the percentage of rural residents
in each state. The second fund involved the same amount of annual appropriations for instruction in trade, home economics, and industrial

subjects, but with no more than 20 percent devoted to home economics. It was apportioned according to population but was clearly aimed
primarily at urban schools, which developed programs in what were
called the industrial arts. In rural areas, shop work was also a part of the
vocational agriculture curriculum.

This support was much more than a subsidy to states, for the act set
up stringent requirements. It appropriated $500,000 in 1918, and $1
million a year after 1921, to train vocational teachers. For agriculture,
this meant new training programs in the land-grant universities and an
informal tie between agriculture teachers and both the agricultural colleges and the Extension Service. The act established and funded a Federal
Board for Vocational Education to carry out studies and to set overall
certification standards for qualifying secondary schools and for teachers.
State boards had to implement such standards. The federal funds paid up
to half the salaries of certified vocational teachers. The state paid the other
half and was responsible for providing the necessary facilities, which
usually meant fully equipped shops. Unlike other schoolteachers, vocational teachers taught for twelve months and, as a result, had the highest
salaries.[10]

During the summer months, the teachers worked with students on
the farm, helping them implement the latest innovations in record keeping, plant and animal breeding, conservation practices, and the use of
fertilizers and pesticides. To be eligible for such agricultural courses, a
student had to live and work on a farm, or at least express an interest in
becoming a farmer. Each student had to carry out farming or farming-
related projects to gain high school credits in what amounted to a summer internship program. In effect, this program was a training ground
for future farmers. Thus, almost every high school with vocational agriculture instruction (this meant only white high schools in most of the
South) sponsored a club called the Future Farmers of America (FFA).
It was organized on both state and national levels, had annual conven-
tions, and awarded prizes for the most successful projects. In most high
schools, boys could take four years of vocational education, amounting to
one-quarter of the requirements for graduation. In my small high school,
all the boys took vocational courses for all four years because there were
no alternatives in the curriculum. The Smith-Hughes program continued
until 1997, when all vocational subsidies moved to the Department of
Education. Its unique culture survives today only in the still-strong FFA,
which includes chapters for both men and women in high schools and
agricultural colleges.

No other federal programs before 1929 matched, in total impact, those related to research, development, and education. Nevertheless, some were vitally important to farmers and were usually developed in response to their effective political lobbying. For instance, more than any other category of user, farmers chafed at locally high, usually noncompetitive railroad rates. In new farm organizations, especially the Grange, they helped gain state rate regulations. When the U.S. Supreme Court invalidated these "Granger" laws, farmers eventually gained much of what they wanted in the Interstate Commerce Act of 1887. The Sherman Anti-Trust Act of 1890, though less of an issue for farmers, prevented some types of industrial monopoly that would raise the cost of what farmers had to buy. Later, when some farmers' cooperatives seemed to restrain trade and thus violate antitrust laws, Congress exempted such cooperatives in 1922. Consumer protection laws, such as the Food and Drug Act and the Meat Inspection Act, both passed in 1906, helped ensure the safety of foods and thus increased the market for agricultural products. Other laws set standards for apples, limited futures sales in cotton, licensed warehouses, regulated stockyards and meatpackers, and set standards for milk. Most important of all, under the Newlands Act (Federal Reclamation Act) of 1902, the federal government assumed responsibility for developing large irrigation projects in sixteen western states. Contrary to early expectations, this meant major governmental subsidies for irrigation works. The major reclamation project dams (such as the Hoover and the Grand Coulee) would eventually provide most of the early electricity for areas west of the Rockies.[11]

CREDIT AND MARKETING

Some of the most important federal programs before 1930 were those involving farm credit. In many regions, the need was great. In much of the post–Civil War South, high-risk and high-interest liens on crops were often necessary to enable farmers and sharecroppers to plant and harvest their crops. For capital expenditures, farmers often borrowed from neighbors or local banks, executing notes or mortgages. Although most of these were for one year, they were normally renewable, but with no guarantee (the federal government prohibited commercial bank loans for more than five-year terms). Some mortgage companies sprang up in largely rural areas. But in times of low crop prices and thus periods of financial stress for farmers, lenders were most likely to call such short-term loans. What farmers needed was a source of long-term loans at the interest rates that prevailed in other industries.

In 1908 President Theodore Roosevelt established the Country Life Commission to study and report on the special problems faced by rural Americans. It documented the lack of an adequate credit source for farmers. For four years, Congress studied and debated various proposals for credit relief, taking many of their ideas from Europe. Most farmers wanted direct loans from the federal government. Rural banks favored privately owned but federally chartered banks. What Congress created was, in effect, a federal reserve system for farmers, modeled in part on the new Federal Reserve System for commercial banks. It tried to create a system that involved farmer cooperation, private investment, and no cost to taxpayers. At the top of the new system were twelve regional Federal Farm Loan Banks, plus some privately chartered joint-stock banks. The original capital ($750,000) for the Farm Loan Banks came from the federal government, which subscribed the original bonds for each bank. These bonds, which were free from federal or local taxes, were then offered to private investors (farmers had to buy bonds valued at 5 percent of their loans, making them, in a sense, part owners of the banks). In most years the system was fully self-financing (if the bonds were not fully subscribed, the government remained the buyer of last resort). Although the federal government had to provide supplemental funding for these banks during the Great Depression, the banks eventually repaid the government's investment in full. These federal banks did not loan directly to farmers; they discounted loans made by local farm credit associations owned by farmers. Loans were issued for terms up to forty years, with interest rates no more than 6 percent. The maximum loan was $10,000 and could not exceed 50 percent of the farm's value.[12]

The Farm Loan Banks aided able farmers (they held about 20 percent of all mortgages by 1929). They did not help less secure farmers, those with much higher credit risks. A second agency, the Federal Intermediate Credit Banks, was created in the aftermath of the severe depression of 1921–1922. However, it had such stringent terms that almost no farmers took advantage of its loans in the 1920s. Also, up to this point, the federal government had not offered any type of production loans to farmers, including money for seed and fertilizer. Farmers with good credit secured such loans from local banks or, in the South, from local merchants (the notorious crop lien system). Finally, this original farm credit system was totally inadequate to deal with the credit crisis that began in 1930. In some cases, the Farm Loan Banks had to foreclose on delinquent borrowers to remain solvent, adding to the number of rural bankruptcies. Thus, new initiatives in farm credit joined a dozen other new agricultural innovations that began in 1930 and climaxed in 1938 (see chapter 3).

None of the governmental programs that matured before 1930, important as they were to farmers, addressed their most insistent demand: the guarantee of a fair price for their crops and livestock. This demand became one of the great policy issues after the collapse of farm prices in 1921 and was perhaps the most controversial political issue of the 1920s. During these policy debates, all possible alternatives were extensively discussed, and every aspect of American agricultural policy, all the way to the present, had some precedent in the many policy proposals and bills considered in the 1920s.

The problem faced by farmers was largely a product of their own success—more food and fiber than the market could absorb. For locally marketed, perishable vegetables and fruit and raw milk, this was still a regional problem, despite improvements in refrigeration and transportation. For the major storable commodities—corn, wheat, cotton, rice, tobacco—the price-determining market was not only national but also international. The foreign demand for cotton was high and reasonably stable before 1930 but extremely volatile for wheat, pork, and beef (with meat production tied to the consumption of corn and other feed grains). By 1925, even though most prices were close to prewar levels, the majority of farmers were becoming increasingly angry at what they considered unfair prices. This was especially true in the major wheat-producing areas, where price volatility was greatest and labor productivity was increasing at the most rapid pace. In the Great Plains, tractors were revolutionizing wheat production by changes in soil preparation, planting, and, above all, harvesting. Wheat farmers would thus be the first to organize and effectively lobby in Congress on behalf of legislation to guarantee farmers a fair price. This led to the creation of a powerful farm bloc, or farm lobby, that still exists today. Of course, the concept of "fair" begged definition, but farmers generally tied the idea of fairness to the ratio between what they had to buy in order to produce a crop and the price they received for it. They based this comparison not on the obviously inflated and abnormal prices during World War I and its immediate aftermath but on the 1909–1914 period, which represented a golden age for farmers. During that period, supply and demand seemed balanced. A fair price would enable farmers to buy what they needed and still retain a fair profit on their annual product. This meant that their prices would be on par with prices in general—hence the term *parity*.[13]

But what governmental policies could ensure such fair prices, and at what cost? In a sense, Americans are still debating these issues. It is clear that most farmers rejected the equilibrating effect of knockdown competition in a wide-open agricultural marketplace. One obvious answer was for farmers to limit production to match demand, thus gaining higher prices. The problem was that no one could come up with a way to control production on a voluntary basis. Any governmental policy that forced production controls on individual farmers faced major constitutional constraints and warred against American values. Besides, for decades, most farmers had vehemently rejected such restrictive policies. Also, almost all prior agricultural policies had worked in the exact opposite direction—to increase production. In all but localized and tightly organized crop areas, such as citrus or grape growers in California, it was almost impossible for farmers to work together as an associated unit. Voluntary production goals made sense only when all farmers agreed to adhere to such limits. Otherwise, noncooperating farmers could take advantage of the price increases made possible by the "sacrificial" reductions of those who joined in solidarity with the group. Thus, it seemed impossible for production controls to work without coercion of some sort, or at least financial incentives to convince farmers to comply. Such incentives came in the 1930s.

Other proposals that tried to chart a way around production controls proved more illusory than practical. They dominated the policy debates of the 1920s. One such policy involved various schemes to sell surplus production abroad at a loss, thus allowing competition to raise domestic prices. One export scheme dominated agricultural policy debates from 1924 to 1929 and involved slightly different versions of the McNary-Haugen bill, named after its congressional sponsors but advocated by George N. Peek, a former farm equipment manufacturer. The plan, despite Peek's claims, was not simple. As farmers came to understand it, the plan was beguiling or even magical, and its details have become a nightmare for college history students. It began with the idea that farmers needed a protected market similar to that of manufacturers, who were protected by tariffs. To accomplish this, Peek wanted the government to establish export corporations that would buy farm commodities at a fixed price determined either by the parity principle (1909–1914 levels) or by the world price plus the rate of tariff on that product (for wheat in the mid-1920s, this was 42 cents a bushel). When world prices fell below the support level, each corporation would have to sell the surplus abroad at a loss. Who would pay for the loss? Not, Peek emphasized, the federal government, but rather the farmers who benefited from the higher domestic prices (of course, consumers, who had to pay those higher prices, were the ultimate losers). The export corporation for wheat, for example, would determine the amount of the loss, and the farmers would pay an equalization fee to the corporation, proportionate to their share of the total crop; that fee would be deducted from the price received by each farmer. This was, in effect, a processing tax. In theory, if only 10 percent of wheat was exported in a given year, farmers would have to pay back only one-tenth of the gain they received from the higher domestic price.[14]

One other proposal that dominated farm policy debates in the 1920s was the concept of cooperative marketing. For the most vehement critics of export bounties (the McNary-Haugen bill), this was an alternative strategy, but for many supporters of bounties, it was a supplementary and possibly complementary effort. The most influential supporter of this approach was Herbert Hoover, first as secretary of commerce and later as president. Hoover believed that the government should not intervene in the private economy to the extent of setting prices. Instead, it should do all it could to help farmers improve prices through a more efficient, up-to-date marketing system. As it was, individual farmers were at a great disadvantage in an economy increasingly dominated by large corporate enterprises. The way of the future was association or cooperation. Farmers had already formed local cooperatives in most parts of the country and could use them to help market their crops; in particular, local or regional cooperatives could join as one large national cooperative for each crop. Congress had passed the Capper-Volstead Act in 1922, which exempted farm cooperatives from existing antitrust laws. The Bureau of Agricultural Economics (BAE), established by the Department of Agriculture in 1922, had organized a small Division of Agricultural Cooperation, and in 1926 Congress passed the Cooperative Marketing Act, which created an expanded Division of Cooperative Marketing in the BAE. Its work, in many ways, anticipated that of the later Farm Board (see chapter 3), which absorbed this division in 1929 (in 1933 it would become part of the Farm Credit Administration).[15]

It was never clear that cooperative marketing could solve the problem of low farm prices, particularly when overproduction was the root cause. Marketing cooperatives could help solve some problems, such as the often large variance between low prices just after harvest and higher ones later on. Cooperatives could store wheat or corn in warehouses or elevators for later sale, thus stabilizing prices over the whole crop season. Cooperatives could also advance loans to farmers at the time of storage, using the crop as security, thus alleviating the problem of short-term credit. Utilizing the excellent statistics compiled by the BAE on prices and markets and on the amount of production needed in future years, cooperatives could at least propose production quotas, but they had no means of enforcing them. Hoover believed that well-informed farmers would voluntarily and unselfishly reduce their acreage to promote the general welfare of all farmers and the nation as a whole.

After President Calvin Coolidge twice vetoed the McNary-Haugen bill, presidential candidate Hoover made a campaign promise to address the farm crisis, perhaps the most critical problem faced by Americans. President Hoover supported and, in June 1929, signed a new farm bill— the Agricultural Marketing Act—by far the most ambitious farm legislation to date. At its heart was federal support of cooperative marketing. But as usual, with Congress pressured from all sides to enact different farm relief measures, this turned out to be a complex omnibus bill. Some of its policy options went well beyond cooperation and well beyond what President Hoover desired. One provision allowed a new Farm Board (independent of the Department of Agriculture) to use a portion of a $500 million revolving fund to purchase farm commodities in times of falling prices to stabilize the market. When implemented in 1930, this led to the creation of two stabilization corporations (one for wheat and one for cotton), which meant that, for the first time, the federal government intervened in the market to guarantee minimum prices for two crops.[16]

From 1930 to 1938, President Hoover and then President Franklin Roosevelt, working with Congress, matured a very complex farm policy whose main outlines are still evident today. In ways not always clear, and with consequences debated by both politicians and economists, these policies helped shape the patterns of development in American agriculture that made it possible for less than 2 percent of workers to feed more than 300 million Americans, with a large amount left over for export.

2. The Traditional Family Farm

A Personal Account

I was born in 1929 on a small, fifty-one-acre farm in Greene County in the valley of east Tennessee. My father had inherited the farm from his father in 1918, when he was eighteen. He borrowed money to build a good all-purpose barn and a small tenant cabin and then rented the farm for several years to a sharecropper. His net income from the farm was minuscule. He lived with a sister and her family and, at the time, had no interest in farming on his own. Instead, he spent a short time shucking corn in Iowa (a great adventure) and then, in the early 1920s, spent almost two years as a hired hand on a rather large farm in southeastern Iowa. He practically became a member of the farmer's family and kept in touch for the rest of his life. Sadly, the owner lost the farm in the Depression. After farmwork, my father spent two years working in automobile factories in Detroit, returning to Tennessee in the summers during retooling time at the factories. But when my parents married in 1928, he decided to try to make it on his own as a farmer. He idealized farming then, and he always would.

My mother, eight years younger than my father, had grown up in the same small community. After high school, she completed one quarter at East Tennessee State Teachers College and had planned on a career in teaching. However, a mix-up in her high school credits forced her to drop out, and she went to work in a rayon factory in nearby Elizabethton. She had grown up on a farm and was not anxious to move back to a small farm and a less-than-adequate house. But in 1928, it seemed that one or two good burley tobacco crops would easily pay off the $500 mortgage my father had obtained to build the existing house and barn, which would then allow him to build a decent house. Then my parents could enjoy the independent life of freehold farmers. My birth did not alter these plans, but the stock market crash that would occur just as I was born (late October 1929) would place their farm at risk and lead to five years of almost desperate poverty. Tobacco prices fell to such low levels in 1932 that our crop scarcely earned enough money to pay the hauling bill to get it to market. For a short time, my family reverted to a near subsistence form of agriculture. We had goods to sell on the market, but at prices too low to cover the cost of production. Everyone in the village suffered from this lack of demand. And it is this pattern of life in the village during the Great Depression that I want to describe.

Profile of a Farming Village

Aided by geological survey maps for 1937, I counted all the farms in my community, as defined by the location of the Bethesda Cumberland Presbyterian Church and the graveyard behind it. The people who lived around this church and attended it (if they attended any church at all) and those who chose to be buried in its cemetery made up a rather cohesive neighborhood. The village was further defined by the upper watershed of a creek and a highway that connected Greeneville with Kingsport, each seventeen miles from the church. Most children attended the two-room schoolhouse in the center of the village, except for those who lived on farms across the nearby Washington County line. Of course, these boundaries are a bit arbitrary, but I identified eighty-two families that lived in this rural farming community in the early 1930s. I included two permanent tenants but did not include a few temporary sharecroppers. With the exception of one absentee owner, all the local farmers knew one another, exchanged goods and labor, and more often than not met in church on Sunday. As a boy, I knew all the local owners and most of the tenants.

My village was in a former mixed farming area, but in the early twentieth century, burley tobacco became the main money crop. For frugal and efficient farmers, the tobacco check that came in December or January was by far the largest monetary income and the source of any savings. But tobacco, a very labor-intensive and challenging crop, took up only a small fraction of the cultivated land. In our area, only a few farmers grew as much as three acres of tobacco; most grew only an acre or two, and the smaller farms grew less than an acre. The government acreage allotments, which began in 1934, were based on farm size and past levels of production. But even before 1934, farmers were restricted in the amount of tobacco they could grow by the labor requirements (at that time, forty-five days of work for each acre of tobacco) and the available space in tobacco or general-purpose barns. The role of tobacco somewhat equalized the incomes of large and midsized farms, for the tobacco crop of a farmer who cultivated or pastured only thirty acres (the amount on our farm) might be the same size as that of a farmer who cultivated sixty acres (close to the maximum cultivated acres on any local farm).

Besides tobacco, all the local farmers grew corn, a necessity for livestock; small grains, primarily wheat, but often oats for horses; and hay crops. I do not know of a single farmer who did not own at least one milk cow, hogs for butchering, and chickens for eggs or broilers. All large and middling farms had at least two horses or mules. No one farmed with oxen, and until the eve of World War II, no one had a tractor. Few had beef-type cattle, although many used beef breeds for their bulls in order to get heavier veal calves from their milk cows. These calves were usually sold at eight weeks of age, a small but dependable source of income. Some farmers had surplus wheat, corn, or hay for sale on local markets. Wives sold eggs, butter, and young chickens to local stores or, in some cases, to selected customers in town. Those farmers with a number of milk cows sold separated cream in the 1920s, but in the early 1930s they all switched to the sale of grade B milk to the Pet Milk Company, which established a factory in Greeneville. Few had more than five or six milk cows. Everyone milked by hand, for no one gained electricity until 1940. By World War II, three or four larger farmers—those with ten to twenty milk cows—qualified for the sale of grade A (or bottled) milk, and dairying competed with tobacco as the most profitable local enterprise.

The Tennessee valley had some very good farmland. Greene County, though joined to the southeast by the Unaka mountain chain (up to 4,800 feet), was largely in the valley of the Nolichucky River or the higher land that stretched north of the river. It had sections of relatively level land. It led the state in burley tobacco production and, by the 1930s, in dairying. But my village was in the uplands that marked the watershed between the Holston and Nolichucky rivers, at an elevation of 1,500 to 1,900 feet. It was hilly in parts, with some land fit only for woods or pasture. But this limestone area had reasonably fertile soil, with a clay subsoil and good drainage, and stretches of only slightly rolling land. It was on the clear, or limestone, branch of Lick Creek, which cut through the whole county from the northeast to the Nolichucky at its western end. The main branch of this creek meandered through a rift-type valley that was almost flat, with some wide meadows that often flooded, but its slate soil was not as fertile as the limestone soil in the uplands. The hills in our village were heavily cultivated and subject to major erosion, particularly on south-sloping hillsides. And despite traditional rotation patterns (corn or tobacco, grain, hay, pasture), much of the land was nutritionally depleted, meaning a dependence on manure and fertilizer. By 1930, few farmers had adopted conservation strategies such as holding dams, terraces, or even contour farming on hillsides. Our creek and its feeders ran red with soil particles after every heavy rain.

When I began my survey of village residents, I was surprised at some of my findings. Everyone at the time defined our community as a farming village, and on census forms, most men listed themselves as farmers, which was about as good a classification as possible. In my high school years during World War II, every boy took vocational agriculture, most for all four years (most girls took home economics). All boys belonged to the Future Farmers of America, and most of us assumed that we would follow our fathers into farming. Actually, only a very small minority of boys would earn a living in agriculture (only one from my graduating class—my future brother-in-law, who would move to a large farm in Pennsylvania). A major economic transition was under way in our area, one that involved changes not only in agriculture but also in manufacturing. We simply did not recognize it, even though many of the changes were happening right under our noses.

Only about fourteen of the seventy family farms in the village were large, by local standards. One was owned by a businessman in Kingsport and farmed by a manager. These farms were all more than 100 acres, with only two or three as large as 200 acres. They had large, mostly two-story, traditional houses and at least two barns (one for livestock and hay, another for tobacco). Most had four horses. All these farms had tenant houses and one tenant or sharecropper. These large farms all had the more expensive horse-drawn farm equipment, including a binder and a grain drill. They depended on custom threshing. Yet because of the small fields and the hilly land, some of the equipment already present on midwestern farms was absent. They still plowed with a two-horse, single-turning plow. Few had riding cultivators. They used fertilizer, but it contained little or no expensive nitrates. Their farm technology largely dated from the late nineteenth century.

I have classified thirty-nine operators as midlevel farmers—the most typical in our community. My father was one of these, and so was the one black man in our community. These farmers owned 40 to 100 acres of land. Their houses tended to be smaller or not as well maintained compared with those on the large farms. They usually had only one barn, which had to house livestock, hay, and tobacco. Most had two horses. The only midlevel farmers who had tenants were a few absentee owners or older, retired farmers. Few could afford binders and thus paid neighbors for the grain harvest. Ownership was stable in the 1920s and 1930s, but not occupations. Only two lost their farms in the Great Depression. Six owners gave up farming during the Depression and moved to near-by cities. Four, including my father, continued to live on the farm and do some farming but took full-time factory jobs. One sold his farm in what amounted to the consolidation of two farms. Today, many of these smaller farms have been further subdivided; some are largely pastureland and are essentially hobby farms. Only one or two are farmed by full-time operators. None of these farms contribute much to the total agricultural product or produce a net income for their owners.

Approximately seventeen other family heads would have listed them-selves as farmers in the 1930 census, but they were marginal farmers at best. They owned less than forty acres, and some as little as ten. They rare-ly had nice homes, but most had at least small barns. Only about half had a team of horses. Most had at least one milk cow, two hogs, and plenty of chickens. Some grew a small patch of tobacco. For all these families, the farm was primarily a subsistence operation, for they had little to sell. None made a decent living from their farms, and almost all did off-farm work, even if nothing more than working as day laborers on larger farms in the area. One man made a fairly good living buying furs from the young boys who trapped and collecting medicinal plants in the summer. One lost a small farm and became dependent on charity. Four were able to procure full-time off-farm employment in the mid-1930s. Two wid-ows had sons who worked off the farm. Four older men, of retirement age, depended on their extended families for income. For all practical purposes, these farms were already outside commercial agriculture. Dur-ing World War II, all able-bodied men and women on such marginal farms were able to get jobs in town. Some kept the old farmstead, but not as an income-producing asset. Many still enjoyed having a few cows or growing a garden. Our community, even in the late 1930s, was on its way to becoming a far-out suburb of industrializing Kingsport, where many people worked at the huge Eastman Kodak plant. With the growth of full-time, off-farm employment, some of these marginal farmers had postwar incomes equal to those of established farmers, resulting in an equalization of local incomes.

Finally, our village had twelve families that did not live on enough land to qualify as a farm. They had lots of one to three acres. Two of these families had sunk to the lowest possible level, both economically and socially. They were, in our unfair language, trash. They had only pickup work and mere shacks for homes. The other ten families owned their homes and had nonfarm employment. They had gardens, hogs, and chickens, but most did not have enough land for a cow. One was the village blacksmith, two others owned grocery stores, two were widows (one with retirement income), one worked in a local veterans' hospital, one clerked in a store in Kingsport, and three worked at the Eastman Kodak plant.

The type of agriculture that was dominant in our village had some advantages in the Great Depression. Only two families lost their farms. We almost lost our farm, but few of the other large or middling farmers had mortgages. Many had small savings accounts in local banks and suffered when a Greeneville bank failed, but their farms were not in jeopardy. The credit economy had not arrived in this part of the county. Elsewhere in the county, where larger river-bottom farms predominated, and in much of the Corn Belt of the Midwest, it was often the largest, most specialized, credit-using farms that went under with the collapse of commodity prices. Had farmers in our area been as "progressive" and gone in debt to buy extra land or possibly a Farmall tractor, they might have suffered the same fate. Without such debt, local farmers suffered a severe loss of income but were able to weather a few bad years. Most had to give up some luxuries. They bought as few products as possible at local country stores and made less frequent trips to Greeneville or Kingsport, where one had to go to find hardware, medications, and dry goods.

By 1930, all the larger farmers had automobiles or trucks, and so did a good share of the midlevel farmers. A few had to give up their automobiles (as did my father). Because of the Depression, self-sufficiency became a necessity. The change was not dramatic, however, because in almost all farming areas in the United States, it was simply not yet profitable to divert labor from the production of goods consumed at home to the production of marketable commodities. Also, farmers used few outside services. For instance, with no phones and the nearest town seventeen miles away, farmers had to be their own veterinarians; most learned how to aid in the birth process and how to use home remedies to treat wounds. Every farmer in my community was a bit of a farrier, too, for they shod their own horses (I think I could still do it), using custom-made horseshoes.

HOME PROVENANCE

Certain subsistence items were so common as to be taken for granted. These included milk cows, chickens, hogs, vegetable gardens, cured meat, canned fruits and vegetables, firewood for cooking and winter warmth, and honey from backyard hives. Having such items reflected rational choices by farmers, for they offered the best real income available for the labor expended. This was generally true nationwide, not just in east Tennessee. As late as 1950, 78 percent of farmers kept chickens (today, less than 1 percent) and 68 percent had milk cows (today, less than 2 percent). Outside arid areas, most still heated their homes with wood. A majority of farms remained without electricity until after World War II.

Yet one qualification is in order. In many cases, it would be more correct to refer to this subsistence side of agriculture as community based rather than farm based. No one farmer grew all the locally produced food products. A degree of specialization had already developed. Thus, local exchange was critical, but it was a type of market involvement that never showed up in farm censuses. During the fall, almost every farmer butchered hogs, but the dates were staggered so that one family would share freshly ground sausage or tenderloin with neighbors one week, and two weeks later would receive reciprocal gifts. Only a few people kept beehives, but everyone had access to cheap honey. Only one local farmer had a mill to squeeze the juice from sorghum and a vat to boil off the excess water to produce sorghum syrup, which we mistakenly called molasses. Others, such as my father, took wagonloads of fresh-cut sorghum stalks to this mill and, in effect, borrowed it to produce syrup. During the summer, only one wife might have extra early tomatoes, which she shared with neighbors. Not all farmers had bountiful orchards, but apples and pears were plentiful because of a surplus on other farms in the area. Swaps of produce balanced out the local accounts, in the same way that farmers exchanged specialized labor. My grandfather, for instance, was an expert in meat cutting and was thus in demand on neighboring farms. Those skilled in carpentry or with mechanical abilities received money or food in exchange for their work. In times of heavy labor demands, farmers pooled their labor to accomplish tasks such as cutting and grading tobacco.

Although horses were, in a sense, the most important animals because they performed the work on the farm, one could not get milk from a horse, and no one considered eating horse meat. In my community, every farmer had cows, hogs, and chickens—the "big three" of farm animals. Even most owners of small lots had at least hogs and chickens, which were still an integral aspect of farm life. Only a minority had sheep, goats, geese, ducks, or turkeys. Of the "big three," the cow was most important. For a yearlong supply of milk, butter, buttermilk, cottage cheese, and yogurt, a family needed two cows. Since a cow goes dry of milk for a few weeks before bearing a calf, calving (and thus breeding) had to be scheduled so that one cow had a calf in the spring and one gave birth in the fall. But during certain times of the year, only a very large family would be able to consume all the milk produced, so even small farmers might sell some butter at the local store. For home use, most local farmers preferred the smaller, higher-butterfat breeds, mostly Jerseys and a few Guernseys, rather than the larger, heavier milkers such as Holsteins or Ayrshires. A good Jersey cow's milk contained 5 percent or more butterfat. The raw milk, before the cream rose to the top, was almost yellow. The cream could be whipped, used in coffee, or made into ice cream. When soured and churned, the whole milk produced an abundance of butter, leaving the buttermilk with small bits of butter floating in it. We used this for cooking and drinking. I still drink a glass of buttermilk each day. It is as good as yogurt for your stomach. My mother molded and sold butter until we began selling raw milk in the mid-1930s. Each family had a distinctive mark or brand on its butter mold, and each wife had to establish the merits of her butter (clean, well colored) before local merchants would buy it. For several years, my grandmother sold butter and cottage cheese (they complemented each other, for they used different parts of the raw milk) to a select group of customers in Greeneville. Each week she took a bus to town, delivering the food in person and receiving well above the market price.

Hogs were almost as important as cows. Of course, one could also get meat from cattle, sheep, or even goats, but it was not easy to preserve these meats. Farmers who butchered steers would sell most of the meat locally. Beef was so rare in our home that we considered it almost a delicacy. Hogs were easy to raise and ate almost anything. They lived for most of the year on leftovers from the kitchen (slop), including extra milk, vegetables, and bits of meat. We also fed them pumpkins from the cornfields in the fall. But before the butchering in November, we fattened them on corn. At hog-killing time we had all the fresh meat we could eat and more, including the liver, tongue, brain, and feet, plus freshly ground and seasoned sausage. We canned sausage and prime loin meat for later use. All the rest—hams, shoulders (Canadian bacon), jowls, sides (bacon)—we trimmed and cured, either in a salt pack or in a mix of sugar and spices (we called this sugar-cured). This cured meat had to last a year and was the major source of animal fat and balanced protein in our diet. No one in our neighborhood smoked the pork, although we all referred to the building where we hung the meat as a smokehouse—a testimony to past practice. Finally, the fat from all the trimmings went into the outdoor cast-iron kettle, where it was boiled and stirred continuously to render the lard, which provided the only shortening we had for the next year. A family of four needed at least two one-year-old hogs weighing more than 300 pounds to meet these annual requirements.

One had to have chickens in order to have eggs. These were indispensable ingredients in cooking as well as for eating (fried, scrambled, poached, boiled, or deviled). In one sense, one could have eggs on the cheap. Chickens, if left loose, could browse and scratch enough food to survive and lay eggs in season. But to have reliable egg production over the year or to grow flavorful fryers or broilers, one needed to care for the chickens. They required grain (cracked corn or purchased feed mixtures), plus gravel and plenty of clean water. They also had to be contained by a fence (to protect them from automobiles and to prevent hens from secreting their eggs), and they needed a dry place to roost (the coop or chicken house) and straw-lined boxes for their eggs. Feeding the chickens and gathering the eggs was a daily chore that was often assigned to children. An annual chore was cleaning out the chicken house and collecting the nitrogen-rich manure for the garden. And, to ensure the heaviest egg production, one performed a task called culling during the winter months. By using one's fingers to measure the spread of the hens' pelvic bones (wide for layers), one could identify which ones were still laying eggs. Layers might be kept for a second year, while those not laying went to market or into the pot. To ensure freshness, local merchants candled all eggs (because they were fertilized, they spoiled quickly), except for those from the best-known and most dependable families.

Chickens were also a much-prized source of meat when I was a boy. Unlike today, when chicken is usually the least expensive meat of all, it was more dear in the interwar years. Chickens required more labor than hogs and, by weight, more than beef cattle. We did not hatch our own chicks in the spring. Instead, like all our neighbors, we ordered them in the mail from any one of dozens of suppliers. Those who were willing to pay more could buy only female chicks, but we almost always went the cheap route and got a mixture. We wanted males for frying or, when older, for broiling. We always ordered the purebred, lighter egg breeds (mostly White Leghorns). The all-purpose, or broiler, breeds were for people who sold the chickens on the market or kept only a few layers for eggs to be used at home. One of the great expectations in the early summer, as the young chickens matured and the males became identifiable, was the first fried chicken of the season. If we ate chickens at other times, they were the older, culled hens that had to be boiled. In the early summer, we jumped the gun and fried young roosters that were scarcely two pounds. They were tender and flavorful, unmatched by any chickens now on the market, even free-range chickens. For two months, we might have fried chicken once a week or whenever company came (always when the preacher ate with us). But chickens were our most costly food, in the sense that we could get a good price for fryers at the local store or in town. We simply could not afford to eat very many of them.

This sketch of home-produced foods can be misleading, because a large number of food and other products were purchased on farms by 1930. It was a very different world from that of a hundred years earlier. Farm families bought sugar, even when they kept bees for honey or boiled sorghum for syrup. Salt, soda, baking powder, oatmeal and dry cereal, dried beans (the largest source of winter protein in most local families), rice, gelatin desserts, pastas, canned beans and salmon, and herbs and spices were all purchased at the local store or from traveling salesmen. In my village, few farmers were still taking their own wheat to mill by 1930; they sold it and bought bags of flour from the store (in part to get the white flour needed for appealing biscuits and cakes). We did take corn to the local mill, however. This old mill, powered by the creek, continued to grind grain and even operate a sawmill until about 1947. Some farmers ground their corn to produce feed for cows, but almost all added high-protein supplements such as cottonseed meal. By the late 1930s, most wives no longer tried to grow peaches (they were often killed by late frosts) and instead bought their fruit from truckloads hauled into the area from South Carolina. In addition to food, farmers bought soap (my grandmother was exceptional, in that she collected lye from wood ashes and fat from table scraps and made her own soap). Most men smoked, and they bought cans of tobacco and rolled their own cigarettes. In the depth of the Depression, a few took tobacco from their curing barns and baked it in an oven to save money. Farmers also bought garden seeds or plants, new varieties of fruit trees, cloth, most clothing, shoes (although they often repaired shoes at home), and furniture.

Most of the burden of providing food fell on the wives and daughters in our community, as in most farming areas. The men prepared the soil for gardens by spreading manure, plowing, and harrowing. For vegetables that took a lot of space—sweet corn, potatoes, squash, pumpkins, and melons—the farmer might interplant vine crops in corn or tobacco fields (a practice now precluded by the use of herbicides and the closer planting of corn), plant sweet corn in allocated rows in the cornfield, or set aside a yard-size patch for potatoes or sweet potatoes. Women or children tended the kitchen gardens and thus most of the vegetables. Farm diets were not always healthy (too much fat, starch, and fried foods; too few green vegetables), but in the summer this was the result of choice, not necessity. Milk and eggs were plentiful year-round. Cured pork provided meat in the winter months. It was possible to store potatoes and sweet potatoes, some varieties of onions, and fall-type apples and pears in root cellars for winter use. Fruits could be dried in drying huts or on rooftops and stored for winter use. Some vegetables, notably cucumbers and beets, could be pickled in apple vinegar, and cabbage could be turned into sauerkraut. By 1930, most vinegar was store bought, although one neighbor still made cider and let it ferment in barrels to turn into vinegar (a passing tradition). Nevertheless, despite refrigerated railroad cars and a reasonably complete highway system, out-of-season foods were generally unavailable in stores, particularly country stores, before the coming of electricity. Thus, the "hungry" time on farms was the early spring, when most stored roots and fruit ran out or rotted, and even the most short-seasoned vegetables (lettuce, radishes) were not yet ready to harvest. It was then that early, mustard-type greens and canned vegetables and fruit were most valued. In the nineteenth century, the invention of vacuum-sealed jars had been a major revolution in food preservation, exceeded only by the advent of frozen foods after World War II. By autumn, every housewife had hundreds of quart and half-gallon jars stored in the basement or root cellar.

What about wild foods? These were no longer a significant food source. This was true generally, not just in my part of the country. Some farmers loved to hunt and occasionally brought home squirrels or rabbits to eat. Most also did some fishing and, when lucky, had fish for dinner. In my area, the deer had been exterminated (they are now back), but except for the elk-hunting areas of the Rockies, game was not a major, let alone dependable, source of meat. The same is true of most nuts, berries, and wild fruit. Children gathered hickory nuts for winter eating, but the meat was small and hard to get at. The American chestnut had been a wonderful source of food in the past, but by 1930, the last of these trees were dying from the blight. Everyone had black walnut trees, with their hard-to-crack nuts. We always cracked enough to make homemade candies, and I loved black walnut ice cream. One widowed aunt, impoverished with three daughters, collected bags of walnuts in the fall, cracked them day after day, and sold the meat for a good price (but not if one counted the difficult labor). During World War II a national market for whole black walnuts led to widespread collecting.

Berries were much more important. I loved to eat wild strawberries, but they were too small and hard to pick to be an important food source. Almost every family had strawberries in the garden and cultivated, not wild, varieties of gooseberries and currants. Many people loved wild raspberries (two types grew locally) and gathered them to make pies or jellies, but they were never a major source of food. My mother planted improved varieties of black-cap raspberries and used them primarily as a dessert. Much more important in the eastern part of the country were wild blackberries. In our normally acid soil, they grew in profusion in pastures or along the fringes of woods. They were free for the picking and lent themselves to the making of pies, jams, and jellies. If families did not avail themselves of blackberries in July and fill jars of them for winter use, they were considered indolent and irresponsible. Thus, picking blackberries was an annual summer ritual, one that I still observe when conditions allow it. I do not remember anyone in our area who cultivated blueberries, but each summer people from the mountains to the south peddled wild blueberries (huckleberries) both in town and along the highway that ran through our village.

HOUSEHOLD PATTERNS

Women's work was never done, particularly on farms. During the busy harvest time, they worked in the fields or barns; at threshing time, they prepared huge meals for the crew. They did at least half the milking, gathered slop for the hogs, and did the planting, cultivating, picking, and preserving of garden vegetables. They had full responsibility for the chickens, including killing them, scalding and removing the feathers, and cutting them into pieces for stewing, baking, or frying. They cooked three hot meals a day, seven days a week. Restaurants were simply unavailable except on those rare trips to town—and then no more than once a month, if one could afford it. Women did the laundry, ironing, and mending of clothing. Most, like my mother, sewed continually. My mother had a Singer sewing machine that she operated by foot pedals. She made almost all the dresses, skirts, blouses, and even suits for herself and my sister, as well as the occasional men's shirts. She pieced quilts and even did less fancy quilting by machine. In some years, she made some extra money by sewing for less skilled neighbors. All these tasks consumed enormous amounts of time and labor, particularly before the advent of electricity. But even in the home, new technologies were slowly reducing labor requirements, although not quite as rapidly as in the field. Nothing as yet rivaled the binder or the tractor.

Before electricity, one could not have running water in the house or barn, except in very rare cases when upland springs or streams allowed gravity to move water into a home. One could, at best, pump water by hand from springs, wells, or cisterns. For a few years when I was a boy, we could not afford a pump and drew water by rope and bucket from a cistern. A few people carried water from a spring some distance away. This was a great burden on wash day (always Monday), when one had to have at lest ten gallons of water. The wash water was heated by fire in an outdoor cast-iron kettle and then moved in buckets to washtubs, where the clothes were soaked and scrubbed on a washboard, rinsed in another tub, wrung out by hand, and hung on outdoor lines. Ironing day followed on Tuesday. Before electricity, flat irons had to be heated on top of woodstoves. At least two irons were required: one heating on the stove, ready to replace the one being used as it slowly lost its heat. Baths were usually on Saturday, in anticipation of church on Sunday. The washtub did double duty as the bathtub. Hot water usually came from a hot water storage bin on the wood-fired kitchen range. For cooking purposes, a tea kettle was always kept hot on top of the stove.

For laundry, change was already under way by 1930. As early as 1900, manufacturers had created crude washing machines: a tub with wheels and an agitator turned by a hand crank, with a hand-cranked, two-roller wringer. By the 1920s, companies had attached small gasoline engines to such washers. Since Maytag was the largest manufacturer, we referred to these as Maytag washers. They were wonderful laborsaving machines. In fact, the first electric washers were identical except for the source of power. In about 1936, as we slowly emerged from the poverty of the Depression, my mother was able to buy her first Maytag. By then, one could hear the chug, chug of Maytag washers throughout the village every Monday morning. Even if automatic washers had been available, farms without electricity and running water could not have used them.

Cooling was a problem in the summer, just as frost-free storage was a problem in the winter. The most widely available solution for both was the root cellar or springhouse. Spring water, at fifty-five degrees in east Tennessee, did not freeze in the winter and cooled milk or melons to the same temperature in the summer. And on a hot day, fifty-five-degree milk seemed cold. A root cellar or a dug-out basement, if mostly underground, would cool to almost the same temperature. This was also the safest place to store canned fruits and vegetables and root crops. But neither the spring-house nor the cellar was cold enough to keep milk or meat for very long without spoilage. The new replacement was the icebox. In the early to mid-1930s, an ice company in Greeneville began sending ice trucks out to our farm. Twice a week, the truck delivered fifty pounds of ice to our icebox, located on the back porch. It cost 10 cents. The ice kept foods and beverages at a temperature close to forty degrees, for relatively safe storage. Those off the main roads were not so lucky. Some did not get ice delivery until after World War II, when the coming of electricity ended the need.

For the few farmers who could afford it, electricity and electrical ap-pliances predated rural electrification. One farm that adjoined ours had installed a generator and a bank of storage batteries by the early 1930s (called a Delco system, after the manufacturer). This family was the envy of the neighborhood. The generator, with a small gasoline motor, charged the batteries at intervals. The main purpose was to provide electric lights, but this family also had an early, direct-current version of a refrigerator and a pump, which enabled them to have an indoor bathroom—the only one for miles in any direction. Also, they had a simple electric radio, while everyone else who could afford it (for us, this was in 1936) had a large, cumbersome radio powered by a set of three batteries; when one battery went out, we had to do without the radio until we could make a trip to town for a replacement. This became easier for us in 1937, when my father bought an automobile. The radio supplemented the hand-cranked phonographs already present in the homes of successful farmers, player pianos in a few homes, and reed organs in several others.

When people first gained electricity (beginning in 1940), they re-ferred to their electric bills as their light bills (some still do). Lighting seemed to be the most important result of electrification for both the house and the barn (electric lights eased the burden of after-dark milking and feeding in the winter). Before electricity, the simplest form of lighting was the kerosene wick lamp, which was used in homes and in the form of mobile lanterns in the barn. But even home lighting was not completely dependent on such lamps, with their yellowish light. Most homes had at least one mantle-type lamp, with small, attached pumps to create the pres-sure needed to force the vaporized kerosene through the mantles. In qual-ity, this white light rivaled that of incandescent bulbs today. Rows of such lamps were suspended from the ceiling in both church and school, but this required a good bit of preparation before any evening event.

Improved roads (our state highway was first paved in about 1936) and the radio helped open up the larger world to farm folks. For some, so did the telephone. For farm villages, the telephone was a reasonably inexpensive new tool. In the 1920s, the men of my village dug holes for poles and stretched copper wires to almost all the homes. A crippled woman served as the operator, with six or eight lines reaching out from the center. These were necessarily party lines, with up to seven or eight families per line and a coded ring for each. The large telephones had batteries, a small magneto, and a crank to provide the low-voltage current needed to send the signals. Until recently, I kept the old telephone my parents acquired in 1928. Our village system had a connection point with neighboring systems, and it was possible to make long-distance calls, although few used this expensive service. For reasons that are not clear, the community gave up the phone system in the early Depression and waited for more than thirty years to get phone lines from countywide systems.

One major task on farms was acquiring the fuel for heat and cooking. One could buy coal, which was relatively inexpensive. The county supplied it for the two large stoves in the schoolhouse. For one year just before the war, my family bought coal for the heating stove but not for cooking. We did not like the odor or the dust. Thus, almost everyone heated and cooked with wood, since almost every farm had extensive woodlots. None of the local houses were insulated or had double-glazed windows, so the heat loss was immense (but the danger from carbon monoxide slight). Also, the stoves and fireplaces would not have worked if the houses had been tight, for outside air had to enter to provide oxygen for combustion and to move air up the chimney. Most of the older homes, including the largest ones, dated from near the turn of the century or earlier and had multiple fireplaces, both downstairs and up. Because no one had time to collect enough wood to supply all these fireplaces in the winter, very few people heated their bedrooms. Instead, they used multiple blankets or quilts (almost all women quilted and sometimes gathered in the afternoons for quilting parties). In the winter months, people tended to congregate in front of a stove or fireplace in the dining room, next to the kitchen, and made a fire in the living room or parlor only when entertaining visitors. Most living rooms had at least one bed as well as chairs, particularly in small houses. Furniture was sparse.

Even with these strategies, a single home could use up to twenty ricks (a stack measuring eight by four feet and eighteen inches deep) in a year— or much more with multiple stoves. Since the kitchen range was heated by a wood fire, firewood was used 365 days a year. Even in the terrible heat of midsummer, wives had to cook the main, noontime meal in the kitchen on a hot range. It could be stifling. Families usually moved to a screened-in porch to eat. Until electricity, there was no relief from the heat—not even overhead or exhaust fans. It must have been worse in the distant past, when houses did not even have screened windows to keep out insects.

Some new technologies helped lessen the burden of cutting so much wood. Dry wood made the best and safest fires, so it was best to cut wood a year before use. Green wood burned more slowly, provided less heat, and could quickly create tar in stovepipes or chimneys, with the risk of fire. The normal tools for procuring wood were the double-bit ax and the two-person crosscut saw. If sharp and well set (to clear itself), such a saw in the hands of skilled operators could cut a foot-thick log in two or three minutes. A good ax could quickly remove the limbs. In a day, two people could cut up to two ricks of wood, but the work was extremely strenuous. By 1930, circular saws pulled by a stationary one- or two-horsepower gasoline engine (one or two cylinders with a large flywheel) could cut up to twenty ricks a day, if supplied with enough small logs. These logs first had to be cut, trimmed, and dragged by horses to the site of the saw. All in all, such saws could reduce by half the work required to gather firewood. When cut into eighteen-inch blocks, the larger slices still had to be split by a single-bit splitting ax. For the cookstove, the pieces of wood could be no larger than two inches in diameter, which meant a lot of splitting. And to keep the wood dry, one needed a shed for it. Most farms had a woodshed that could hold a winter's supply. Finally, carrying wood into the house was a daily chore almost always assigned to boys.

All these tasks may suggest that farm life was full of drudgery. I am sure some women felt that way at times. Men liked to boast of hard work and lament poor returns, even as they valued the independence of farm-work (being their own boss, they said). But the total work output, mea-sured not only by time but also by intensity of effort, was significantly less than a forty-hour workweek in a factory. The farm was confining because of the care of livestock, but the work was usually leisurely. As a matter of strict religious principles, no one worked on Sunday except for necessary chores and the cooking of meals. Laundry on Monday was hard work, but the amount was slight by contemporary standards (one set of sheets per bed, two or three towels, and only one set of soiled clothes for each per-son). A broom was used for most housecleaning. Children usually washed and dried the dishes. My mother, despite all she accomplished, always took an hour after lunch for a nap. My father never worked, or required me to work, beyond noon on Saturday, except during the harvest. Almost everyone attended church twice on Sunday, and everyone went to socials at the schoolhouse. Most important, life became easier with each passing year. The great concern at the turn of the century about the deficiencies of country life had led to significant improvements in almost all the cited problems—poor schools, low incomes, inadequate roads, poor health care, and an absence of cultural resources, such as libraries. Note that these concerns involved country people, whatever their occupation, not just farm families. The two poorest families in my community did not farm.

By 1935, Tennessee had finally extended the school year to eight months for elementary schools and nine months for the new, mandated high schools. Around 1930, Greene County and neighboring Washington County began to run school buses to most rural communities. This meant that, for the first time, graduates of Bethesda Elementary School could live at home and attend high school. Such attendance was not yet mandated, and too many local students stopped their education at the eighth grade. But within the next decade, high school became the accepted norm for almost all families, and by then, a few students even went on to college. My two-room school was, in a sense, primitive. By necessity, it was heated by stoves, had outdoor privies, and, until the end of the 1930s, had no well for water (we carried it in buckets from a spring two-tenths of a mile away; our water bucket and dipper are visible at the far right of figure 7). Except for one year, the school had two teachers, each

Figure 7. Bethesda Elementary School, September 1936 (I am in the second row, fourth from the left).

with some college preparation. Each one had to teach four grades, and most were generally capable and conscientious and did a very good job. Their students who moved on to Fall Branch High School, in Washington County, were class valedictorians in four out of eight years.

Health care was a challenge. No physician lived close to our village. The nearest hospitals and pharmacies were in Greeneville and Kingsport, each seventeen miles away. Some people died, needlessly, because they did not go to a physician or a hospital until it was too late, if at all. My county did not have an adequate public health program until the New Deal period. By the time I entered school in 1935, a visiting public health nurse came at least once a year and began mandatory vaccinations for smallpox, typhoid, and diphtheria. No treatment existed for other childhood diseases such as measles, whooping cough, and mumps. Epidemics of each occurred periodically, with a few deaths. But with better roads and automobiles (for most families), trips to town and physicians' visits (when necessary) slowly became the norm. By 1940, most farm families in our area had the same access to health care as did working families in cities. Even before this, we had one dentist in nearby Fall Branch.

My family was lucky to live on a state highway. In the early 1930s, it was already an all-season road, with plenty of gravel and daily grading. In about 1936, the county paved it, eliminating the stifling dust we suffered each summer. By then, a Trailways Bus line began service between Greeneville and Kingsport, allowing us to flag a bus in front of our house. Also, a local company ran shift buses to Kingsport. Thus, our area was not isolated, and we had had daily rural free delivery since soon after the turn of the century. Before 1937, we ordered most clothing from Sears Roebuck and garden seed from Henry Fields in Shenandoah, Iowa. As noted earlier, we even bought chicks by mail. Most families in our area received a daily newspaper by mail, either the *Greeneville Sun* or the *Knoxville Journal*. As our family income improved in the mid-1930s, we also subscribed to two farm journals, one or two women's magazines (for my mother), and, in some years, periodicals such as *Life, Look,* or the *Saturday Evening Post*. Most influential in my life was the *Progressive Farmer*, a now famous magazine for southern farmers that tried to disseminate the most successful new farming methods to ordinary farmers. Thus, the cultural isolation was over for most of us, but some families who lived off the highway still suffered winter roads that were all but impassable.

The purchase of our first automobile in 1937 changed our lives dramatically. Before this, we rarely went to town. My mother would take the bus about once a month for necessary items such as hardware, medicines, clothing, and schoolbooks (this last was an intimidating burden for poor families, since the state mandated the purchase of books but did not pay for them). Now, with a car, the whole family could go to town almost weekly. For the first time, we attended movies. By age twelve, I was able to get a library card at the Carnegie Library in Greeneville. We went to town often enough for me to check out books for two-week intervals. Starved for reading material, I now read all the time—or whenever my parents or work schedules permitted. By then, I did not feel that living on a farm amounted to cultural deprivation.

In retrospect, the 1930s marked a major, enduring transition in American agriculture. Life on the farm in 1930 was closer to that of 1830 than 1960, so rapid were the changes already under way. Unbeknownst to us at the time, much that we took for granted would quickly become antiquated. Soon there would be no horses, or none of the tools and skills that accompanied a horse-based agriculture. Soon it would become clear that our farming community did not have the soil quality or the economies of scale to compete in commercial agriculture. Soon agriculture ceased to be a major source of income for all but a handful of farmers.

Another change was almost as radical. Slowly, often without realizing the significance, local farmers began to buy more food in town and grow less on the farm. For those who did not sell milk, it was soon uneconomical to keep a cow. After World War II, the efficiency of production in almost every specialized area of agriculture and the efficiencies in the processing and marketing of foods made it cheaper to buy almost any type of food than to grow one's own. Also, in a change I sometimes regret, fresh fruits and vegetables became available year-round, not just in season. Even cultivating certain garden vegetables was no longer cost-effective, if one considered the amount of labor involved. Yet, either out of habit or as a hobby, most older families continued to grow vegetable gardens. Some still do. They were positive that store-bought vegetables lacked in flavor. But for new generations, the loss of flavor seemed a small price to pay for the convenience, particularly after the advent of frozen foods. Besides, commercial farmers have to commit all their time to the products they sell on the market. They are so efficient that Americans pay only about 8 percent of their income for food consumed at home (the growing amount consumed in restaurants adds another 5 percent). How we arrived at this point is the subject of the following chapters.

3. A New Deal for Agriculture, 1930–1938

The Great Depression necessitated new farm policies. Otherwise, up to half of America's farmers would have suffered bankruptcy (one-fourth did), and the overall depression, destructive as it was, would have been much more severe. Over eight years, during two presidential administrations and four Congresses, the federal government, responding to a large array of interest groups and competing policy alternatives, matured a complex body of laws and administrative agencies to gain what everyone hoped would be fair and stable prices for almost all major agricultural products. Details have changed through the years, but aspects of every policy option undertaken in the 1930s have endured until the present, providing the political constraints and opportunities that allowed American agriculture to remain the most productive, and food prices to remain the lowest as a percentage of total spending, in the world. The human costs of this transition were enormous.

The new farm legislation extended the debates of the 1920s, with no completely new options either debated or accepted. Two presidents—Herbert Hoover and Franklin D. Roosevelt—were the final arbiters of this legislation, but more often than not, the details were thrashed out in the Congress. Hoover was much involved in the details of farm legislation and had a personal agenda for agricultural reform, one that mainly involved governmental support for cooperative marketing. He fought but often failed to achieve his goals, and at times he reluctantly accepted bills that did not reflect his values or meet his objectives. Perhaps more insistently than any president in history, he tried to shape agricultural policy. Even though he lost many of his battles, Hoover's detailed input into farm policy meant that the eventual farm program as implemented by 1938 had his imprint on it.

Roosevelt was just as committed to the welfare of American farmers as Hoover. If anything, he had a deeper personal identification with their problems than did Hoover. What he did not have, and would never develop, was a coherent set of values and principles to guide his policy inputs. He was open to many alternatives, sometimes even contradictory ones. Thus, whereas Hoover tried to focus and constrain farm policy, emphatically rejecting many of the competing policy proposals of the 1920s, Roosevelt was, at one time or another, open to almost all strategies. For example, New Deal programs would include acreage or production controls, export bounties, cooperative support, food aid, monetary inflation, land retirement and limited resettlement of displaced farmers, subsidized commodity loans, marketing contracts, and relief and rehabilitation programs for the poorest farmers.

First Fruits: Hoover's Farm Board

A new farm program began in 1929 with the Agricultural Marketing Act, by far the most significant legislation of that year. Its implementing agency was the Farm Board, the first of a long line of agricultural agencies with more than a dozen different names that tried to aid in the marketing of farm products. The Farm Board was originally supposed to use its $500 million revolving fund to create and integrate farm marketing cooperatives. The huge wheat crop in the summer of 1929 and the attendant decline in wheat prices, followed by the stock market crash in October, presaged a more general crash in the prices of several farm commodities in 1930. Although it was not clear at the time, the general economy, led by a volatile agricultural sector, was moving into the worst depression in American history.

In 1930 Congress faced pressing demands for new policies to stabilize or raise farm prices. The new Farm Board had to use some of the seemingly minor provisions of the Agricultural Marketing Act to stabilize prices. Over the next three years, or until the summer of 1933, debates over farm policy raged as farm prices continued to fall. A series of crises, each worse than the one before, was part of a collapsing world economy, with the nadir for farmers coming in 1932 and early 1933. In the midst of this gloom, the beleaguered Farm Board did all it could to keep prices as high as possible for major farm commodities, and for short periods it was able to stabilize domestic prices above world levels. But it faced an impossible task and ended up spending nearly all its $500 million in a failed effort. In the process, it provided a rehearsal for several of the agricultural programs of the New Deal.[1]

The stated aims of the Agricultural Marketing Act were to decrease speculation in agriculture, prevent waste and inefficiency in marketing, encourage farmers to join cooperative associations and thus gain leverage in the market, and aid farmers in preventing and controlling surpluses. The large revolving fund was intended to be used not as a direct subsidy to agriculture but as a source of low-interest loans to cooperatives, loans that they would repay in due time. If regional cooperatives in each major crop could combine into one national association, they would gain as much market power as large corporations. Farm Board loans could enable cooperatives to acquire new facilities, such as grain elevators or cotton warehouses; make advances to farmers at planting time; provide up-to-date information about market conditions; and stabilize prices throughout each crop cycle. To achieve this stability, the cooperatives could, on the recommendation of a commodity advisory committee for each crop, set up corporations to buy and store crops, with the goal being more profits, not subsidies to farmers.

To guide what Hoover saw as an impending agricultural revolution, the Farm Board was headed by Alexander Legge. Seven additional directors represented the major agricultural industries: corn, wheat, tobacco, livestock, cotton, eastern fruits and vegetables, and California fruit. To a large extent, this board fulfilled its primary goal. Most of its staff and employees worked to bolster cooperatives and to help those cooperatives combine into one centralized association for each crop, or something close to a cartel (images of the later National Recovery Administration). The benefits, as everyone conceded, would come over the next several years. But almost as soon as the organizational work was under way, farm prices collapsed. This crisis soon absorbed the board's attention. Without the Depression, the Farm Board might have achieved more of Hoover's idealistic, associational goals.

Although no one initially conceived of the Farm Board as a tool for price manipulation or control by the federal government, it soon backed into that role. One central goal was price stabilization. The idea was simple: Farmers could rarely afford to hold their corn, wheat, cotton, or tobacco off the market during the harvest season, when prices were lowest. They needed the income to plant the next season's crops. Few farmers had the facilities to store crops. Processors, either cooperatives or investor-owned companies, did have the facilities and often gained higher profits by holding crops until the off-season. If farmers joined in local and regional cooperatives, and these in turn held the stock of national marketing associations, farmers could stabilize prices over the year and gain the greatest returns for their efforts. Local cooperatives could offer low-interest production loans to farmers, based on the security of their stored crops, or what amounted to a crop loan system owned and managed by the farmers themselves. With what seemed little risk, cooperatives could make crop loans for up to 80 or 90 percent of market price at the time of the loan. In most cases, the price would be higher at the time of optimal sale, with the gains going to the farmer (legally, the farmer repaid the loan and interest at this point and regained possession of the market crop).

Thus, beginning in 1930, crop loans, at first largely for wheat, became an enduring and often misunderstood aspect of farm marketing. If farmers had the opportunity to loan their crops to a cooperative at a set loan rate (a percentage of the market price), they knew that, at the end of the season, they would almost inevitably gain a certain return on their crops. This loan rate would become, as everyone expected, a floor price, at least in the short term. Unlike the nonrecourse loans offered today by the Commodity Credit Corporation, these were normal loans, in the sense that the farmers would have to repay the full amount of the loans, even if prices dropped below the loan rate. This seemed a very unlikely prospect, unless the whole agricultural economy collapsed. But that is exactly what happened in 1930, and the Farm Board had no statutory authority to absorb the losses suffered by cooperatives. Farmers were thus in debt to the very cooperatives they owned. The cooperatives, in turn, many of which had just been established with loans from the revolving fund, were in debt to the Farm Board. And in 1930, neither was in a position to meet its payments. Although not bound by law to rescue its debtors, the Farm Board was under intense pressure from farmers and from farm-state congressmen to do just that. Using the only legislative remedies it had, the board began the long and complicated federal effort to control agricultural prices and, in the process, provide large subsidies to farmers.

As noted earlier, wheat farmers suffered declining prices during the summer of 1929, just before the stock market crash. The problem was an oversupply of wheat caused by a rapid expansion of wheat production in both the United States and other major wheat-producing countries, plus a record harvest in 1928 and a large carryover of surplus wheat in the United States. This increase reflected the rapid shift to tractors and combines, which increased the number of acres planted each year in the Great Plains. However, bad weather in 1929 had caused a decrease in wheat production, which seemed to portend rising prices after the harvest. This seemed the perfect opportunity for the new Farm Board to do its thing and, in the process, gain the support of more farmers for these new marketing strategies. Wheat prices were at $1.30 a bushel in July 1929—not the price farmers had expected, but not alarmingly low. Wheat prices dropped from August to October, causing farmers to despair. The Farm Board believed that short-term circumstances had caused the decline and that prices would soon recover. Thus, it took a gamble, one that had momentous consequences. It loaned funds to enable wheat cooperatives to support crop loans at 90 percent of the market price, or $1.18 a bushel. The cooperatives did this by supplementing more cautious loans from private banks or the government-chartered Intermediate Credit Banks. To be granted these loans, farmers had to market through the cooperatives. Recognizing the long-term problem of overproduction, the Farm Board also launched a major educational campaign to get farmers to reduce their wheat acreage by 10 percent in 1930 and by 25 percent over the next three years, a campaign that completely failed.

At least temporarily, the fixed loans stabilized wheat prices at or slightly above the loan rate, even in November and December 1929, after the stock market collapse. But by January 1930, the export market for wheat had dried up, and the cooperatives soon had millions of bushels of unsold wheat. Prices fell well below the loan rate. This meant that the cooperatives had to collect on the crop loans from farmers, or at least force farmers to meet new margin requirements—a situation similar to the margin calls in the stock market. Most beleaguered farmers and many local co-ops were in no position to survive such a credit crunch. It seemed as if the disaster of 1921 might repeat itself. The Farm Board felt responsible and sought a way to rescue its clients. It chose an option from the Agricultural Marketing Act. On February 11, 1930, it organized a Grain Stabilization Corporation that began buying wheat, both from farmers who still had it in storage and from local cooperatives, at a set price of $1.18, the loan rate.

This first stabilization effort clearly aided farmers, but otherwise, it was a disaster. The Farm Board did not want to reward private firms that held wheat, so it bought only in local markets. Since the price paid was soon 20 cents or more above world prices, private elevators and export merchants began shipping wheat back to local markets, leading the Farm

Board to purchase only from its own cooperatives. Many private dealers lost tons of money, either in disposing of wheat or on futures contracts. The whole market system entered a period of near chaos, which worsened an already dismal export market. Soon the Grain Stabilization Corporation owned millions of bushels of wheat that it could not sell at anything close to its cost. But, to be fair, no one had anticipated what was happening in the world economy. Everyone had expected an early recovery of normal wheat prices and thus no long-term losses on the wheat bought. Farmers loved the Farm Board, but private grain dealers hated it. A governmental corporation using tax funds to stabilize farm prices at a noncompetitive level was, from their perspective, a form of socialism or communism. Hoover became their devil, but in most respects, he defended his Farm Board and, as usual, condemned the business interests that placed private profit above what was clearly in the public interest. By March, it was clear that the Grain Stabilization Corporation could not maintain the $1.18 support price, and it stopped buying wheat.

The Farm Board entered the market again in late spring and summer, but this time its loan rate was at only 80 percent of world prices. This temporarily stabilized wheat prices at about 81 cents. Once again, the board had rescued farmers from disaster and seemed to avert a widespread panic. Since this effort took place in the open market, it aided private buyers of grain as well as farmers. But the good news did not last for long. Despite a reduction in wheat production during the severe drought in the summer of 1930, prices collapsed again in 1931. By this time, the Farm Board had all but exhausted its revolving fund and did not resume stabilization purchases. By then, the stabilization corporation owned three-fourths of the wheat carried over into the summer of 1931. Without any authority to force farmers to reduce production, the board had to leave farmers to the mercy of the market. By 1932, wheat prices had fallen to 48 cents.

Beyond wheat, the Farm Board established only two other stabilization corporations—one for cotton and a regional one for California grapes. Cotton, unlike wheat, was completely dependent on exports, with more than half the cotton produced in the 1920s shipped abroad. Despite some competition from Egypt and India, southern cotton clothed the world. Unlike wheat, the collapse of cotton prices did not precede the stock market crash, but quickly followed it. Cotton, as a nonfood crop, has an elastic demand, since people can postpone the purchase of textiles, and governments, short on foreign credit, can impose barriers to its import. Early in 1930 the Farm Board refused to set up a stabilization corporation for cotton but used crop loans to try to keep prices above 16 cents a pound. It also tried, without success, to get cotton farmers to reduce the acres planted in 1930, although a summer drought in the lower Mississippi valley did cut production below 1929 levels. Nevertheless, cotton prices continued to fall in the spring of 1930, leading the Farm Board to form its second stabilization corporation in May. The buying of 1929 cotton carryovers stabilized prices for while, to the benefit of farmers. But supplies of unsold cotton mounted in warehouses, and the Farm Board stopped buying cotton by the summer of 1930. Cotton prices continued to decline, reaching 7 cents a pound in December. The Cotton Stabilization Corporation then resumed cotton purchases, leading to a brief rise in prices early in 1931, but by that time the Farm Board did not have the funds to make further purchases. By then, everyone realized that the central problem for both cotton and wheat, in a rapidly shrinking world market, was overproduction. Pleas to farmers did not work. In 1931, for example, cotton farmers had a banner year, and by the fall of 1931 the carryover of cotton roughly equaled the annual production.

What could the Farm Board do to get farmers to reduce production in any crop? Several ideas floated about, but in only one case did the Farm Board succeed. This occurred in the small, regional grape industry of California. The Farm Board's creation in 1929 seemed to offer a lifeline to what had once been one of the most successful farm cooperatives in the United States, the Sun Maid Raisin Growers Association. In the heady days just after World War I, it had overexpanded, offered generous production loans to its members, and accumulated large supplies of raisins. In 1929 it faced bankruptcy. Equally desperate was the California Vineyardists' Association, which represented the grape growers who, after Prohibition, sold fresh grapes and grape juice to a weak market. In August 1929 this association proposed that the Farm Board create a federal Grape Stabilization Corporation to represent all the grape producers in California, whether they produced raisins, juice, or fresh grapes. In 1929 the Farm Board offered a loan but not a stabilization corporation. The loan paid part of the costs of making production loans to the Sun Maid cooperative and even financed the repurchase of Sun Maid bonds, but falling prices in 1930 doomed such rescue efforts.

In March 1930 Charles Teague, vice chairman of the Farm Board and the representative of California fruit growers, offered all the grape growers a unique package, one that skirted the limits of Farm Board authority. The Farm Board would provide credit for a new Grape Control Board—something like a hybrid between a marketing cooperative and a stabilization corporation. To gain Farm Board funding, the Grape Control Board would have to sign growers, responsible for 85 percent of all grape production, to ten-year contracts that required them to sell their grapes only through cooperatives affiliated with the control board. They also had to pay the control board an annual stabilization fee, collected at the time of sale, of $1.50 on each ton of fresh grapes and $5.25 on each ton of raisins, with the predicted $2.5 million yield used to buy up an anticipated 250,000 tons of surplus grapes. These fees were closely related to an equalization fee proposed in the McNary-Haugen bills, or an early version of the processing tax established by the first Agricultural Adjustment Act in 1933. All growers also had to pay $2 a ton to help retire the debts of Sun Maid. Most interesting, Teague argued that the best way to deal with the projected 250,000 tons of surplus grapes was to pay growers not to pick the grapes and leave them to rot in the fields (for this, they would eventually receive at least $10 a ton at a time when the average sale price was $59, a price made possible because of the reduced supply).

Despite problems, this stabilization plan was the only Farm Board initiative that dealt head-on with the problem of overproduction. It took a huge, fervent local campaign to get enough growers to sign the contracts. Certifying the grapes withheld from production proved difficult, and some cheating undercut the plan for fresh grapes. Most critical, the surplus for the year soared to 770,000 tons. The control board held these grapes off the market to support higher prices, creating a potential problem for the following year. Fortunately (or unfortunately), poor weather and insect infestation almost ruined the 1931 crop, at least temporarily eliminating the problem of surpluses and low prices. Before this crop loss occurred, a local committee of the control board had decided to offer growers $5 an acre and $10 a ton (based on past levels of production) to destroy grapevines and not replant them for six years. The Farm Board liked the idea, but its attorneys pointed out that it had no statutory authority to make loans that curtailed production. In many ways, the Grape Control Board in California presaged later New Deal initiatives.

By 1932, the Farm Board had used up its revolving fund and was almost helpless to prevent a further decline in farm prices. Chairman Legge had resigned in March 1931, and the vice chair had resigned in June. By then, the board was barely able to keep the cooperatives afloat. At the same time, the federal farm credit agencies faced bankruptcy and had to tighten, not loosen, loan requirements. It was now clear that the only solution to the farmers' woes had to involve some means of controlling the supply of products offered on the market. Many proposals competed, and many of these received support from the Hoover administration. For example, the president supported credit subsidies that would allow foreign governments to buy surplus crops; this led to limited sales to Germany, China, and Brazil, but hardly enough to influence domestic prices. He also supported concerted efforts to get southern farmers to plow under every other row of cotton in 1931, while the Farm Board tried to get state governors to convince farmers to plow under every third row. Some economists in the Wheat Belt urged the Farm Board to pay farmers $2 an acre to withhold land from production, but this would have required new legislation from Congress, and the costs would have been prohibitive. Some southern states tried to take responsibility for production controls. Mississippi, Texas, and Oklahoma proposed mandated cuts of 40 percent of their cotton acreage if three-fourths of all cotton states would pass comparable laws. None did, and no one clarified how to enforce such limits. Huey Long in Louisiana had the most radical proposal. He suggested that all cotton farmers cancel the 1932 crop, which would almost exactly wipe out the carryover and thus allow competition to restore fair prices. Long wanted all the states to levy stiff fines on any farmer who planted a crop. His plan was unanimously approved in the Louisiana legislature, but it would go into effect only if all the other cotton states agreed to do the same. Only one other state passed a similar bill.

Maturing a New Farm Program

During the tense legislative battles of 1932, Hoover increasingly linked the farm problem to that of overall economic recovery, which in turn depended on critical international cooperation. He supported credit expansion at home and higher tariffs to protect the domestic economy, approved funds to rescue the Farm Loan Banks, backed a new loan program for cooperatives, and approved loans for agriculture from the new Reconstruction Finance Corporation (RFC). But by then, he was even more fearful of foreign dumping, direct relief, and, above all, several plans to gain monetary inflation (the old free silver issue), which would jeopardize his efforts to achieve some form of international economic cooperation. As the presidential election campaign approached, he seemed more rigid than ever, less willing to consider initiatives he might well have accepted in 1930. Burned by the failure of stabilization efforts, he wanted to eliminate the stabilization provision in the Agricultural Marketing Act. But in one area—production loans—Hoover reluctantly had to accept relief for farmers.

Both the very cautious Intermediate Credit Banks and the cooperatives supported by the Farm Board made loans to farmers, but not high-risk loans to farmers who had exhausted all their income from the sale of their crops and thus needed money for seed and fertilizer to begin the next crop cycle. By 1931, the proceeds from harvested crops rarely covered the cost of production, so most farmers were in need of this short-term credit. Traditionally, they had secured such credit from neighbors, merchants, or local banks by signing a lien on the crop to be planted, but with falling prices and often unpaid mortgage debts, farmers were a high credit risk, and lending institutions were leery of making such loans. Desperate farmers pressured their congressmen to get the federal government to grant what everyone called "seed loans." It had done so in the past after natural disasters, and after the drought of 1930, it had appropriated more funds than Hoover wanted for such loans as a form of drought relief. In 1932 a reluctant Hoover approved the use of RFC funds to offer seed loans to all farmers who could qualify, and he even more reluctantly approved a smaller amount to provide food and other necessities for destitute farmers. Many farmers were never able to repay these loans, which thus amounted to a type of farm relief. The Department of Agriculture administered this loan program, with local Extension Service agents screening the applicants. Hoover also approved a new short-term lending program—twelve regional agricultural credit corporations—that had less stringent requirements than the Intermediate Credit Banks but did not make seed loans.

By the election of 1932, it was obvious that when the new Congress convened in 1933 it would have to approve a newly packaged farm marketing program. The plight of farmers and the clout of the farm bloc ensured this. At the same time, it was obvious that, after a decade of controversy, the new agricultural legislation would include no novel provisions. Every conceivable approach to the crisis in farm prices had been represented in bills before Congress or in the actions of the Farm Board and farm credit agencies. Several credit initiatives would be consolidated in a new Farm Credit Administration. Several proposals to lease or purchase submarginal land came to fruition in the Land Policy Section of the new Agricultural Adjustment Administration (AAA). The equalization fees in the McNary-Haugen bills would become a processing tax on foods and fibers. To the extent possible, the AAA would subsidize the sale of farm commodities abroad. The crop loan advances of the Farm Board and the work of the stabilization corporations would continue under one large stabilization and crop lending agency, the Commodity Credit Corporation (CCC), which was unique only by virtue of its greater generosity to farmers. An amendment to the Agricultural Adjustment Act made it possible for the government to leave the gold standard and, for a year, dramatically inflate the currency, which was what the most radical farm reformers had wanted since the time of the free silver campaign. Seed loans would be expanded into a major relief and rehabilitation effort for less competitive farmers. Finally, the localized crop-reduction efforts of the Grape Control Board in California would expand nationwide into a new domestic allotment system, which became the centerpiece of the New Deal farm program.

Except among grape growers, the great unsolved problem in agriculture in 1933 was overproduction. All groups suffered during the Depression, but farmers suffered more than producers in any other economic sector. In 1932 farm incomes were less than half what they had been in 1929, land values had dropped by 40 percent, and the per capita income of farm families was down by at least two-thirds. It was in this context that Congress conducted an extended debate about new farm policies. None of the contending factions prevailed, leaving the new Roosevelt administration to deal with these unresolved issues in 1933. However, the debates did give rise to a slowly evolving plan for a domestic allotment system that repackaged several older initiatives.

The two most popular agricultural reform proposals in the 1920s had involved either cooperative marketing or foreign dumping of American surpluses, or some combination of the two. Advocates of cooperative marketing hoped to persuade farmers to cut production voluntarily whenever surpluses accumulated. Advocates of dumping hoped that it would be possible to maintain tariff-protected domestic prices well above world prices without forcing farmers to curtail production. But to gain this benefit, farmers would have to use some type of fee or tax on their domestic sales to pay for losses on crops that they had to dump on foreign markets. This led to the equalization fees in the various McNary-Haugen bills. This two-price system remained popular among farmers as late as 1932, even though it was an unrealistic solution during a period of rampant economic nationalism and drastic drops in foreign trade. As early as 1925, some agricultural economists had recommended that governmental programs, to ensure higher domestic prices, under the protection of tariff walls, should be conditional on contractual cuts in production on the part of farmers (the idea adopted in 1930 by California grape growers). This led some to advocate a system of domestic allotments, whatever the means of allocating and enforcing them, but it was clear that any such system would have to be, in some sense, voluntary. Also up for grabs was the type of compensation needed to gain farmers' support for what they had long resisted—bureaucratic control over their production. By 1932, almost everyone agreed that some kind of processing tax, levied on price-supported crops, would pay for the domestic allotment system. The problem in 1932 and 1933 involved what form these various measures would take.[2]

During the congressional debates of 1932, a well-planned crusade for production control through domestic allotments gained the support of a majority of farm organizations. Before the 1932 election, both presidential candidates offered rather vague commitments to a new farm program. The most influential backers of allotments, including farm economists Milburn L. Wilson and Mordecai Ezekiel, had hoped to gain Hoover's support and thus make domestic allotments a part of the Farm Board. They failed. Hoover, who had earlier supported some system of voluntary acreage controls, decided against production controls. But Wilson and Ezekiel gained the backing of Rexford Tugwell, a member of Roosevelt's so-called brain trust, and Henry A. Wallace, who would later become secretary of agriculture. They thus had direct access to Roosevelt, whose farm proposals were general and vague but clearly supportive of some type of domestic allotments. But Roosevelt was also open to several other possibilities for dealing with the agricultural crisis.

This meant that a new farm bill would be at the core of recovery legislation passed in the new Democratic Congress after March 1933. This Agricultural Adjustment Act proved to be one of the most controversial New Deal bills. It was enacted on May 12 after two months of contentious debate, well after some farmers had planted their crops. The law as passed was among the most complex ever enacted in the United States, and among the most influential. Some parts of it are still operative. It was an omnibus bill, in part because of a series of amendments that packed the bill with almost every conceivable method of aiding farmers. It went well beyond agricultural marketing. It included a mortgage relief measure, and its inflation amendment (the Thomas Amendment) revived the free silver issue. It made silver as well as gold acceptable for foreign debts, authorized the issuance of silver notes, and, most critical, allowed the president to alter the gold content of the dollar. Thus the Agricultural Adjustment Act became the means of implementing a completely new monetary policy, and the United States periodically changed the dollar price of gold as a means of inflating domestic prices and, not incidentally, supplementing the tariff by blocking foreign imports. The cheap dollar lowered the price of American goods in foreign markets and raised the dollar price of imported goods, at least until foreign governments adopted similar tactics.[3]

THE AGRICULTURAL ADJUSTMENT ACT OF 1933

The 1933 act established a new agency within the Department of Agriculture, the Agricultural Adjustment Administration, which replaced the Farm Board. Its first head was George Peek (of McNary-Haugen fame). The AAA had the authority to use multiple strategies to boost agricultural prices and dispose of surpluses, many carried over from the Farm Board. The main financing provision of the act—a processing tax on all supported commodities—was, in effect, a sales tax on food. The revenue from this provision could be used to pay export subsidies, but because difficulties in foreign trade led to limited exports, dumping would have little effect on farm prices. In later years, various forms of subsidized exports would become a major aspect of American agricultural policy.

The two principal programs developed by the AAA were marketing agreements between farmers and processors and the domestic allotment system for selected commodities. Only the largest commodities (cotton, wheat, and hogs in 1933, with tobacco and corn added in 1934) participated in the allotment system, and they have received the most attention. They also provided most of the lobbying support that led to the Agricultural Adjustment Act. But the vast majority of crops, including almost all perishable fruits and vegetables, as well as the large beef cattle industry, either rejected allotments or were never considered for participation because of the inherent attributes of their marketing. In 1933 growers of all but one specific variety of tobacco (used for cigar wrappers) agreed to accept domestic allotments for 1934 and signed contracts by the end of 1933. For 1933 the AAA negotiated a marketing agreement with seven tobacco companies that raised tobacco prices to parity levels, with the tobacco companies absorbing the costs of increased prices.

For seven other largely regional crops, most without export markets, the only way to establish minimum prices was to negotiate agreements with processors. For these regional crops, particularly fruits and vegetables, the AAA licensed food processing companies and helped negotiate marketing agreements. In many cases, processors united to offer parity, or close to parity, for a set quantity of produce. Such price-fixing agreements were possible because of provisions in the National Industrial Recovery Act of 1933 that suspended antitrust laws to allow the negotiation of

cartel-like industrial codes. In the case of food processors, the bill assigned code-making responsibility to the Department of Agriculture, not
to the National Recovery Administration. This allowed processors, such
as tobacco companies, to use fair competition rules or codes to justify
unified prices far above former competitive levels. In some cases, farmers
agreed to destroy, or not harvest, the portion of a crop that exceeded the
amount called for in the marketing agreement. Because of constitutional
concerns about industrial codes, most of these marketing agreements ex-
pired by 1935. New legislation allowed their replacement with marketing orders, which became the mode used to support milk prices.[4]

The act's other major provision, the domestic allotment system, was
the most publicized and most controversial program of the AAA. It lasted
only three years, but it set enduring precedents. The complexity of the
program, including different challenges and different controversies for
different crops, defies any brief explication. The idea was that farmers of
selected crops would contract to cut production by a designated percentage. The expectation was that, over time, this curtailment of production
would raise commodity prices to parity level (the parity principle was
written into the act). Meantime, those farmers who contracted to reduce their production would receive compensatory payments from the
AAA. The amount of these payments would be set each year to bring the
farmers' total income up to the parity level. However, the amount of the
payment was based not on actual production during the crop year but
on average production in earlier years (usually a five-year average). This
allowed farmers to get at least a part of their payments at the beginning of
the crop cycle. Also, when droughts ravaged the Midwest from 1933 to
1936, farmers whose wheat and corn crops failed still received payments,
or a type of crop insurance. A processing tax (really a sales tax) was as-
sessed on each commodity in the program to cover the costs. Since the
program was completely voluntary, nonparticipating farmers gained a
possible benefit from reduced surpluses, but they received no payments.

Such a program seemed, to many critics, almost impossible to administer. How could the AAA set quotas for each farmer? How could it verify
farmers' claims about acres cultivated or past yields and thus fairly assign
quotas? Above all, how could it verify the fulfillment of the contracted reductions? It seemed that nothing less than a huge bureaucracy would be
able to monitor the work of millions of individual farmers. The solution
was to have the farmers administer their own program. For any given
crop, all the farmers who signed contracts made up what could be called
an association. Working at first through county extension agents, the AAA

required these farmers to join local groups (in townships or county districts) and elect from their number a local committee—the lowest administrative level of the AAA. The farmers in each county also elected a three-member county committee. In this early system, a given county could have multiple committees tied to individual crops (for example, one committee for wheat, and another for tobacco). The local committees determined the quota for each farmer, and the county committees determined the amount of production allocated to each area. Above the county committees were state committees. The AAA provided the necessary clerical support for the program at the county level, where most of the action took place. There would soon be AAA offices in every farming county in America.

In theory, the local committee signed farmers to contracts, determined quotas, certified the accuracy of reports, and verified full compliance with contracted reductions. The AAA paid local committeemen for the time they spent visiting individual farms and checking on compliance. But in time, most of the actual work devolved to the clerical staff in the county offices, and even though, legally, the committees made all policy decisions, the full-time staff played a major role in guiding the committees and communicating policies to individual farmers. These local AAA offices became the point of contact between almost all farmers and the national government. Local committeemen retained the obligation, or the right, to monitor acreage allotments, but in almost all cases they hired other people to do this work. I surveyed the tobacco acreage in my district of Greene County for four years.

The maxim "from small acorns great oaks grow" was certainly applicable to the original assignment of domestic allotments. The quota established for each farmer participating in the program became known as the base acreage. The local committees, usually made up of the more successful or prominent farmers, used the two guidelines prescribed by the AAA—the amount of cultivated acres and past production records—to establish this base. For example, if a given farmer cultivated fifty acres and in the past five years had grown an average of four acres of tobacco, his base might be three acres; a farmer with half as many acres and half as much production would have a base of one and a half acres (the goal being an overall reduction of acreage to match production with demand). Although sometimes guided by extension agents, and reflective of local class and racial norms, the local committees generally followed these guidelines and gained strong support from most farm operators.

Totally unanticipated by anyone at the time, this base acreage became the foundation of an enduring aristocracy. Similar to primogeniture and entail laws in Europe, this procedure attached to land a special and valuable privilege. For many basic crops, success depended on gaining land with an established base. As late as 2007, farm payments of various types would be tied to base acres or base marketing quotas, and the local committees were still the arbiters of the annual reallocation of these privileges. Clearly, those who gained a base in 1933 had a head start that endured generation after generation. And when they sold the land, the base shifted to the new owner.

The first year of the AAA was full of both confusion and controversy. Its most notorious action in 1933 led to the plowing under of part of a growing cotton crop and the slaughtering of pigs and sows. In both cases, the effort led to only a slight reduction in surpluses, but it quickly put money into the hands of the farmers who reduced their cotton and hog production. Except for very young pigs, which were used to make grease and fertilizer, the kill actually added 100,000 pounds of pork to government relief efforts. Since corn farmers were not under contract for 1933 (they received stabilizing crop loans in the fall of that year and signed acreage reduction contracts in 1934), the effort to improve pork prices was an indirect way of aiding corn farmers, many of whom fed their corn to hogs. Most wheat farmers contracted for a 15 percent reduction in the 1934 wheat crop, which turned out to be unnecessary because of a severe drought. In fact, the searing heat and droughts of the mid-1930s finally eliminated the surpluses carried over from the old Farm Board. Beef cattle owners voted against production controls, and poultry owners, who were widely dispersed on almost all farms, were not part of the domestic allotment system.

The scandal over hogs led to a major and enduring food aid program. This was new, with the exception of some Farm Board wheat distributed through the Red Cross in 1932. I suspect that direct government purchases of food and cotton during the Depression had a greater impact on agricultural prices than the acreage control programs did. In response to the bad publicity over dead hogs, the Roosevelt administration chartered the Federal Surplus Relief Corporation to buy and distribute all types of surplus commodities, almost three-fourths of which would be livestock. The AAA continued to buy hogs until the summer of 1934, and during that drought-ridden season it also bought 8 million beef cattle. An appropriation of $75 million in September 1933 allowed the purchase, through the AAA, of processed milk products and several types of fruits and vegetables, some of which were distributed through the Federal Emergency Relief Administration (FERA). To some extent, the Federal Surplus Relief Corporation's stress on food needs did not match the surplus removal goals of the AAA.

In 1935 that tension ended when the program shifted from an independent corporation to the newly created Federal Surplus Commodities Corporation, located in the Department of Agriculture. From then on, the commodities purchased and distributed clearly reflected the goals of the AAA—surplus disposal and higher farm prices. Section 32 of a 1935 amendment to the Agricultural Adjustment Act provided that one-third of all customs receipts would be used in support of agriculture—to facilitate exports, encourage domestic consumption, and reestablish farmers' purchasing power. This became the major source of funds for food aid of all types and would remain so until the Food Stamp program in the Kennedy administration (a few food stamps were also issued during the New Deal). The Federal Surplus Commodities Corporation used every possible means to dispose of surplus food, giving it to charitable institutions, county welfare agencies, schools, and needy families. Boxcars full of food were unloaded for county welfare workers to distribute. I remember in 1937 when trucks rolled up to our rural school, delivering dozens of boxes and bags of food. We had no school lunch program, so we ate apples and oranges and endless amounts of prunes. We reluctantly drank some of the sour canned grapefruit juice at school and carried home rice and beans and canned foods, even though some of our homes had no pressing need for them. The "capture" of food aid by the Department of Agriculture proved politically invaluable in future years. It created a second constituency for the department and its programs and often helped deflect criticism from its commodity programs, as it did during debates on a 2007 farm bill.[5]

The voluntary nature of the allotment system caused problems. It worked well for corn and wheat, for it allowed many small farmers, who largely consumed those grains on the farm, to opt out of the program. Not so for the two largest nonfood crops—cotton and tobacco. Strictly speaking, neither fit any "domestic" allotment system, since they remained predominantly export crops. Noncontracting farmers could take advantage of higher prices without reducing acreage, undercutting its purpose. Thus, in two related acts (the Kerr Tobacco Act and the Bankhead Cotton Act), Congress added new teeth to cotton and tobacco marketing in 1934. Under these laws, if two-thirds of farmers voted in favor of production restrictions, anyone who tried to sell a crop not grown on the allocated acres, or any crops sold by farmers who did not accept the acreage restriction, incurred an all but prohibitive tax. In effect, the acreage limitations turned into a universal form of control. Such coercive rules were constitutional only because the farmers had to vote for such a program. From 1934 to the end of all tobacco price supports in 2004, burley tobacco growers voted for these tight controls every year, and light fire-cured growers in all but one year.

In the midst of the herculean efforts to set up the domestic allotment system and make payments to millions of farmers, the AAA struggled internally over its policies. Peek never liked production controls; he sparred with Secretary Wallace over his authority and did all he could to boost export subsidies. Roosevelt supported Wallace and thus had to ease Peek out of the AAA in late 1933. By then, another battle raged because of demands made by appointees in both a legal division and a division on consumer affairs and opposition to their demands by the directors of the main marketing divisions, who reflected the interests of commercial farmers. The lawyers, generally from urban areas and more politically radical, wanted to use the marketing agreements to gain more control over the processing businesses and to ensure that payments to southern farmers did not discriminate against tenants and sharecroppers (who were often African American) either by displacing them or by not sharing the payments with them. This battle raged through 1934 and 1935, until Wallace fired the radicals.

OTHER NEW DEAL FARM PROGRAMS

The AAA was not the only governmental agency to aid farmers in the Depression. A reorganized credit system for farmers was the most important of the non-AAA reforms. It included the Emergency Farm Mortgage Act, which provided $200 million for rescue loans that allowed farmers up to five additional years to repay their mortgages and set a maximum of 4.5 percent interest on old and new loans. It also provided $2 billion worth of guaranteed bonds to enable the Farm Loan Banks to refinance loans and thus prevent massive foreclosures. In time, the banks would be able to retire all these bonds, with no loss to the federal government.

In the Farm Credit Act of 1933, the Roosevelt administration reorganized the array of credit institutions inherited from the Hoover administration. The new Farm Credit Administration took over the seed loans from the Department of Agriculture and absorbed the local agricultural credit associations that had offered some of these loans. The act also established a new Central Bank for Cooperatives, which assumed what little was left of the Farm Board's revolving fund. Twelve regional Production Credit Corporations received the necessary funds to revive the 4,000 farm loan associations, which actually did the lending to farmers.

Although not recognized at the time, the most important new credit institution was the Commodity Credit Corporation. It was chartered under the RFC in 1933, with an appropriation of $25 million to be used to aid farmers. In effect, it took over the crop loans made by cooperatives under the old Farm Board. It was at first an independent federal corporation, but it worked through the Department of Agriculture. The CCC loaned money on the basis of stored crops and at a set loan rate, which continued to be a type of minimum price for farm commodities. It was an important tool for stabilizing prices in late 1933, when farm prices softened and both southern and midwestern farmers were up in arms. As in the 1920s, farmers could redeem the loans and, after paying the interest charges, sell the crop at the market price. Alternatively, they could forfeit the loans without penalty, for these were nonrecourse loans (farmers did not have to pay for any losses). If the CCC profited from the eventual sale, as it often did in the early 1930s, it had to share the gain with the farmers.[6]

One much-discussed but heretofore unimplemented method of reducing agricultural surpluses was the federal lease or purchase of farmlands, thus removing them from production. In a small way, the New Deal embraced this strategy. In 1933, under the authority of the National Industrial Recovery Act, the Public Works Administration committed $25 million for the purchase of submarginal land. FERA did the purchasing, but a new Land Policy Section in the AAA, headed by farm economist Lewis C. Gray, actually selected the farmland. FERA, working through Rural Rehabilitation Corporations in each state, was responsible for resettling the displaced farmers. The original grant expired in the summer of 1935, but by then Gray had used most of it to option or buy almost 6 million acres. A large part of this purchased land became the basis of a state park system. However, the withdrawal of this relatively unproductive land did not have a major impact on agricultural production; in fact, it accounted for less than a third of the funds used by the Bureau of Reclamation to develop new irrigated lands in the West. By 1935, Gray sought an additional $50 million to expand the program but never received nearly that much. He also tried to coordinate land purchases by several governmental agencies (for expanded national forests or parks or for dam construction), but only a part of this involved farmland. Yet, this small start set enduring precedents, and land leasing became an important technique of production control after World War II (this includes the present Conservation Reserve).[7]

As one would expect, this land purchase program, and the insufficient effort to resettle displaced farmers, faced concerted resistance from both farmers and local governments. The program was part of a much more ambitious effort by economists in the Department of Agriculture to establish a more rational use of land in the United States, or what they referred to as land-use planning. For them, the domestic allotment plan was a temporary measure; a planned allocation of land resources was the long-term goal. In spite of excellent surveys of American farmland, estimates of the optimal use of land in each region of the country, and even the formation of local planning committees in most rural counties, nothing substantial came from the effort, which Congress eliminated during World War II. Such planning implied the massive purchase of farmland, the leasing of land, or some type of zoning to control private land use. Congress was not about to grant such authority, even if it were constitutional. Also lurking within this type of planning was the acceptance by farm economists that modern agriculture, with its rapid technological advances, would displace a large proportion of existing farmers, although in 1935 no one anticipated the magnitude of this displacement in the decades after 1945.[8]

Another aspect of New Deal agricultural policy involved help for the losers in agriculture. This affected the increasingly noncompetitive small farmers and the large number of tenants and sharecroppers who, by the very effect of New Deal agricultural policies, would lose their farms or their jobs. For two years a battle raged in the AAA over the plight of sharecroppers in the South, who often did not share equally in AAA payments and, if African American, had no role in the local committees. Advocates for these losers lost the administrative battles, and the AAA became an agency that supported successful or, in its own language, "progressive" farmers. But this does not mean that those who developed the AAA programs were insensitive to the losers. Often, the opposite was true. Milburn L. Wilson, the most important architect of the domestic allotment system, supported programs to aid people at the bottom. He personally headed a small subsistence homesteads program that built communities that combined subsistence plots with part-time industrial employment. Implicit in this program was the fact that large numbers of rural people had no future in commercial agriculture. But the care of displaced or unemployed farmers, including those resettled under the land purchase program, was basically a relief effort. It did not help revive a vital agricultural sector. Efforts within the early AAA to force landlords to maintain all their tenants or sharecroppers failed. Many of these were already redundant.[9]

FERA, through its state-based Rural Rehabilitation Corporations, tried to bring relief to destitute farm families. It had various strategies, including direct aid, but it tried to go beyond the dole by offering small rehabilitation loans to less than creditworthy farmers, some of whom were never able to repay. These loans resembled the seed loans of the 1920s, but they covered a range of needs beyond seed or fertilizer. Perhaps more important, but more intrusive, was the detailed supervision of the farming operations of FERA's clients. The Rural Rehabilitation Corporations in many states also tried more ambitious and experimental programs, including the establishment of resettlement communities on repurchased farmland. Because farm families lived on small plots, and community cooperative enterprises were rarely profitable, these clients were never able to repay even half the cost of such communities. In 1935 both the rehabilitation efforts and the community building moved to a new Resettlement Administration headed by Rexford Tugwell. In 1936, after the resignation of the controversial Tugwell, the Resettlement Administration shifted to the Department of Agriculture and became the Farm Security Administration. It assumed one new task—to supervise a program that involved low-interest loans to carefully screened young tenant farmers to allow them to buy farms. These efforts to improve the lot of small or marginal farmers and to help as many as possible succeed in agriculture worked at cross-purposes with the AAA and its efforts to reduce surpluses. At the end of World War II, Congress abolished the Farm Security Administration; stopped funding its more daring programs, including a few fully cooperative or communal colonies; and converted its successor, the Farmers' Home Administration, into largely a lending agency that continued to focus on small farmers.

Two other agencies, with no connection to the Department of Agriculture or to the AAA, provided vital benefits to farmers. The first was the Tennessee Valley Authority (TVA). It had its origins in the Muscle Shoals project begun during World War I in northern Alabama. That project's primary mission was to construct what became the Wilson Dam and to use part of its electrical output to manufacture synthetic nitrates for munitions. After the war Congress fought intense battles over the fate of this project, with some factions advocating the sale of electrical energy. But southern farmers yearned for inexpensive nitrate fertilizers and wanted fertilizer production to be one goal of the project, whether it remained public or became private. Thus, the act that created the TVA provided for a fertilizer facility at Muscle Shoals. This became the United States' first major research and development center for fertilizer. The TVA not only developed new fertilizers but also established experimental farms in the valley to teach farmers how best to use fertilizer. The TVA thus augmented the already large research and development programs in the Department of Agriculture.[10]

At the same time, the TVA helped form rural cooperatives to distribute cheap electrical energy. Its subsidized loans to such cooperatives demonstrated one way to bring electricity to farmers who lived far from urban centers. In 1936, to extend such efforts beyond the Tennessee Valley, Congress established the Rural Electrification Administration, which funded rural electrical cooperatives all over the country. By World War II, its effort to electrify all American farms was less than halfway completed, but by less than a decade after the war, electrical lines had reached all but the most remote farms in the United States, amounting to a revolution that was almost as important as the adoption of tractors.

SOIL CONSERVATION AND THE AGRICULTURAL ADJUSTMENT ACT OF 1938

Soil conservation became a national obsession in the 1930s, the decade of the great dust storms. The South had suffered the most from water erosion, with red, denuded hillsides and deep gullies widespread in the Piedmont and upper South. The leading advocate of erosion control was Hugh H. Bennett, who headed a small Soil Erosion Service in the Department of the Interior in the early 1930s. His service used its funds to demonstrate erosion control techniques, such as contours and terraces, on selected farms. Fortunately, Franklin Roosevelt wanted to be an even greater conservation president than his cousin Theodore. He thus supported the Soil Conservation Act of 1935, which established the Soil Conservation Service to replace the Soil Erosion Service. It would later move to the Department of Agriculture and was recently renamed the Natural Resources Conservation Service. The Soil Conservation Service was a research and demonstration agency; it did surveys, issued educational literature, set up local demonstrations, developed and leased tools and machinery to farmers, and proposed local legislation. With state cooperation, it set up more than 3,000 soil demonstration districts, most at the county level. After 1954, it would oversee a major watershed protection program.[11]

Most of the funding for erosion control came from other federal departments or from the states. The Civilian Conservation Corps funded and built many of the local projects, some of which were on private farms. But the first major direct subsidy for farmers came in another 1935 law, the first Soil Conservation and Domestic Allotment Act. This act provided for grants to states to support erosion control and other conservation measures on farms. The aid was available only to those farmers who participated in the production control programs. This was a powerful inducement for farmers to sign contracts with the AAA.

The original domestic allotment plan did not guarantee any level of prices for farm products. The payments to farmers, usually made before harvest time, were based on average yields over the prior five years. Also, to the despair of many economists in the AAA, nothing in the plan required farmers to improve their farm practices or protect the land from erosion. This lack was remedied in part by the 1935 Soil Conservation and Domestic Allotment Act. In an unexpected way, this act also came in handy in early 1936 when the Supreme Court, in the famous *Butler* case, declared the processing tax unconstitutional on two grounds. First, it was not a true tax but merely a way to take money from one class of citizens and give it to another. Second, the government used the so-called tax to coerce the behavior of farmers and regulate agricultural production, a power reserved to the states. This claim was based on the prevailing and narrow interpretation of the interstate commerce clause. Contrary to most textbook accounts, the Court annulled only one small section of the Agricultural Adjustment Act, but it was a crucial section, for the processing tax funded all the payments to farmers and all the export subsidies used to dump surpluses abroad. The Court's decision did not affect any marketing agreements that did not involve a processing tax; nor did it preclude land purchases, emergency mortgage relief, food aid, inflationary measures, or a small production control program for cane and beet sugar that was not funded by the tax. Most important, it did not do away with the AAA and its state, county, and community committees.

By January 1936, many farmers had already received payments for the 1936 crop; others looked forward to promised payments. Facing intense pressure from farmers and farm organizations, Congress appropriated funds to fulfill these contracts. This took care of the constitutional objection to an illegal tax. But it did not dispose of the Court's objection to federally mandated production controls, an issue at the heart of a series of Supreme Court decisions involving the reach of the interstate commerce clause. To solve this problem, Congress, in only two months'

time, approved a hastily drafted second Soil Conservation and Domestic
Allotment Act. To circumvent the Court's interpretation of the interstate
commerce clause, the act provided that the states could, if they so desired, request conservation funds from the federal government; establish
state agencies to receive such funds; and, consistent with rules set by the
Department of Agriculture, distribute these funds as payments to farmers
who contracted to accept production controls and to implement certain
conservation practices on their farms. Of course, all the states so requested, and each set up an agency (these varied) to receive and account for
the funds. In every case, the states distributed the funds through the state,
county, and local AAA committees.

In many ways, this constituted an improvement over the early allotment system. Since it now funded payments to farmers from general tax
revenue, it ended the regressive sales tax on food. It required farmers,
each year, to sign a conservation plan under which they agreed to cut the
production of soil-depleting crops (those already under production controls) and replace them with soil-conserving crops or other conservation
uses. For most farmers, this requirement was not onerous, for they could
simply convert land withdrawn from crop production into legumes or
pasture, plant trees on eroded hillsides, or build ponds to hold runoff
water. Most farmers were scarcely aware that anything had changed, but
the adoption of conservation plans had an effect over time, for farmers
had to become aware of soil problems and new ways of controlling erosion. Since conservation practices were farmwide and not tied to any one
crop, the AAA in 1936 changed its organization from divisions tied to
individual crops to one based on regional divisions. It thus worked out
a single contract with each farmer for one or more crops. The new act
was supposed to remain in effect for only two years, but Congress kept
extending it. In fact, the Soil Conservation and Domestic Allotment Act of
1936, though often amended, remained the basis of most federal conservation subsidies for farmers until 1985.[12]

The final major agricultural policy change came in 1938, with the
second Agricultural Adjustment Act. Though large and complex, it generally sought to consolidate farm policy. It did not abolish the first Agricultural Adjustment Act but amended it, nor did it cancel the payments
made under the Soil Conservation and Domestic Allotment Act of 1936.
For five crops—corn, wheat, rice, cotton, and tobacco—all involved in
interstate commerce, it established a new, formalized, and enduring nonrecourse loan plan. Decisions by the Supreme Court in 1937 so broad-
ened the reach of the interstate commerce clause as to make this second

Agricultural Adjustment Act possible. Only farmers who accepted production controls were eligible for such loans. If two-thirds of the farmers
in each of these crops voted in favor of production controls, all farmers in
the main production regions for that crop had to either accept marketing
quotas or pay a large penalty. In effect, this act extended the compulsory
features of the tobacco and cotton programs established by special legislation in 1934. The act allowed, but did not mandate, additional parity payments to farmers if the secretary of agriculture desired them and
Congress appropriated the funds.

In good crop years, with surpluses and falling prices, most farmers
loaned their crops to the CCC and never redeemed the loans during the
nine months they covered. If the CCC profited on future sales of forfeited
commodities, it had to distribute the gains to the farmers who had forfeited their loans. This was a good deal for farmers in any eventuality.
Most farmers never understood the technicalities of the loan program
and simply believed that, at times, the government bought their crops. In
1939 and 1940 the new loan program led to soaring surpluses of corn
and wheat, with growing storage costs. Only the high demand during
World War II solved this surplus problem. The new loan plan, and thus
price supports for basic, storable commodities, would remain in effect
(with many modifications) until 1996, and in a sense, it still continues
today (see chapter 6). Farmers also continued to receive payments for
conservation practices, in addition to these new price supports.

In the Agricultural Adjustment Act of 1938, Congress stressed the
benefits of crop storage, or what Henry Wallace had long celebrated as an
"ever normal granary." But one suspects that this was a political cover for
the CCC and its nonrecourse loans. Since 1954, only adverse weather
or unexpected increases in foreign demand have temporarily eased the
problems of overproduction. The 1938 act also tried to spare a few
farmers from some of the insecurities of floods, droughts, hail, or disease epidemics by including the first quite limited federal crop insurance
program.[13]

This completes the survey of major agricultural legislation from 1930
to 1938. In no period of American history has the federal government
undertaken so many initiatives or inaugurated so many programs to aid
one economic sector. Farmers received payments for cutting production
and subsidies to carry out necessary conservation practices; they received
price supports for five basic commodities and crop insurance as a form
of disaster relief. In fact, the sheer number of new programs still confuses

most historians, just as they confused even the legislators who approved them and the farmers who benefited from them. The long-term effects of these programs continue to invite controversy. One effect, however, was clear. Although farmers led the way into the Great Depression, those eligible for the new farm commodity programs were the first producers to regain anything close to pre-1929 prices. Prices rose to parity level in 1935 and exceeded that level in 1937, meaning only that the prices farmers paid for inputs were equal to or less than crop prices. Because of the overall effect of the continuing depression, farm incomes remained below 1929 levels.

In spite of the confusing array of programs, one factor remained constant, and this is significant. All production controls, all payments for compliance, and all price support programs were based on levels of production. This would remain true from the 1930s to the present. Large farmers accepted the same percentage decrease in acres or marketing quotas as small farmers did, and all payments were based on units of past production. Even policies keyed to farm income rather than commodity prices were based on prior levels of production and thus helped sustain the gap between large and small farmers. Production controls made it more difficult for small farmers to compete with larger ones, and larger and more efficient farmers gained the greatest benefits from farm policies. In the long run, the most enduring benefits of price-raising subsidies were an increase in the value of farmland and an even greater importance for base acres. These results made entry into farming more and more expensive.[14] One long-term effect of this product-based system was a tendency for small, less competitive farmers to leave agriculture, often selling their land to more commercially successful neighbors. At the same time, the large and expanding Department of Agriculture, despite internal battles, continued to cater to its prime constituency—the most affluent and capable farmers. Programs that responded to the needs of small farmers, or to the declining number of tenants, were unable to compete effectively for funds and, in the case of the Farm Security Administration, succumbed in the face of congressional hostility.

4. World War II and Its Aftermath

A Family Report

Like World War I, the Second World War meant higher prices for almost all farm products. A high demand for food crops, based largely on events in Europe, quickly used up all the surpluses held by the Commodity Credit Corporation. The acreage controls and price supports of the 1938 Agricultural Adjustment Act were largely unneeded. Enforced acreage controls remained only for tobacco. Agricultural prices remained at or above parity, and net farm income soared. Although the federal government did not place controls on farm prices, it enforced rigid price controls on all consumer goods. Food rationing prevented severe shortages. These controls indirectly limited what farmers received and prevented a speculative bubble in rural America. Thus, the value of farmland rose only moderately during the war. Yet the war brought enormous change to rural America and reinforced trends already under way that would allow a steady and rapid growth in farm productivity and, more important in the long run, opportunities for off-farm employment for members of farm households.

WARTIME CHANGES IN MY VILLAGE

I was twelve years old in 1941, old enough to participate in most farmwork. I helped milk the four or five cows that allowed our family to sell milk in Greeneville, our only assured biweekly income from the farm. I helped set out tobacco plants in May, hoed and wormed the growing plants, and participated in the cutting and barning of tobacco in August.

By the next year, I was able to drive horses and guide a three-foot cultivator through the rows of corn and tobacco. By 1943, I was almost grown and considered myself a full hand on the farm, and thus I often worked for other farmers for the standard $1 a day (the standard did not go up during the war). In September 1943 I entered high school and began four years of course work in vocational agriculture. In 1944, as my first farm project to meet these course requirements, I rented our one acre of tobacco from my father and took responsibility for the three or four acres of corn we grew for use on the farm. I also worked, as a matter of course, in making hay for winter feed for our cows. Briefly, I was a farmer and looked forward to a career in agriculture.

Since my father worked at Eastman Kodak in Kingsport, he had little time for extra farmwork. During the war he often had to work overtime, so my mother and I largely ran the farm, with my father lending a hand at harvest time. My father had taken his factory job in 1936, but for three years he continued to run the farm on a reduced scale. He grew the full tobacco allotment, as did all the local farmers with off-farm employment. By 1937, we grew only enough corn to feed our chickens and hogs, and by then we had stopped growing wheat. In 1937, with a sharp downturn in the economy, Eastman Kodak placed most employees on part-time hours, which meant more time for my father to work on the farm. But in 1939, with full employment restored and a new house to live in, my father sold his horses and largely gave up farmwork. For the next two years, we had a sharecropper living in our old house. He and his family did all the farmwork except for milking the cows, which my mother would not relinquish. But the farm was not large enough to provide a decent income for the renter, and when the war began he found outside work. This meant that I had an opportunity to take over much of the farmwork and earn some good money. By the time I graduated from high school in May 1947, I had slightly more than $2,000 in the bank, almost all from my share of the tobacco crop. It may not seem like very much today, but at the time it was exceptional (comparable to at least $22,000 today). I used it, plus what I earned doing summer work, to pay my way through college.

An observer would not have noticed many overt changes in the local farm economy during the war. One or two large farmers had bought tractors just before the war. A few lucky farmers along the state highway (including our family) gained electricity in 1940. This was a tremendous boon for my mother, both in our home and in the routine of milking the cows and cooling the milk to strain into cans for the daily milk truck.

But the vast majority of local farmers did not get electricity until after the war (nationally, only 35 percent of farms were electrified by 1941). Since food was so important to the war effort, some young men on larger farms received deferments, but those who did were often stigmatized. Few applied. One who did was a high school student who helped his widowed mother operate a large farm. He was essential to keeping the farm productive, since extra farm labor was almost nonexistent. High school students were released from class to help pick beans on one local farm. Most village sharecroppers quit farming and took defense jobs. Almost none of them returned to farming, and sharecropping virtually came to an end with the war.

Despite the number of men in the military and those who flocked to city jobs (reducing local farm labor by at least half), the level of farm production in our village did not decline during the war. In fact, I suspect it went up slightly, along with prices and net income. How could this be? First, the remaining farmworkers, both men and women, worked more hours. The amount of redundant farm labor on small farms was now clear. In 1942, with no sharecropper, my father went back to farming until I took over two years later. He bought one horse, even though two were normally needed for most farm tasks. He found a rare one-horse wagon and a small one-horse turning plow. We made do with this for tobacco and a small field of corn. For the main plowing, we hired a neighbor who had a tractor. But the point is, we maintained about the same level of output, and with my father's factory job, our family income was higher than ever. The annual income from the farm—primarily from tobacco, milk, and foods consumed at home—amounted to less than $1,200. That might be enough for a family of four to survive on, but it was far less than what most families expected in the way of consumption. Nonetheless, the farm income added to the outside wages, and the value of the farm steadily increased. Eventually, this part-time farming—or, in time, hobby farming—would become the dominant form of agriculture in our village.

New tools played only a small role in sustaining local agricultural production in the war years. The government allowed a limited number of tractors to be produced during the war, and a few local farmers bought small tractors and the new tools for them, particularly Ford-Ferguson tractors. Because of their power and speed, tractors increased productivity, but their greatest impact resulted from their displacement of horses (which had consumed up to a fourth of the product) and the gradual shift to fossil fuel as the source of horsepower. Agricultural research was beginning to have a major impact, with almost all local farmers converting to hybrid corn. Kentucky and Tennessee experiment stations continued to develop new and more productive varieties of burley tobacco. The Tennessee Valley Authority (TVA) had demonstrated the benefits of new fertilizer formulas on county farms, and wartime incomes allowed farmers to use more balanced fertilizers (nitrogen was the critical new ingredient) on their crops, particularly tobacco. Hybrid corn varieties, aided by the nitrogen in fertilizers, raised yields from about twenty-five to more than forty bushels an acre. When grown only for farm use, corn required both less work and less acreage, leaving more land for hay and pasture. Nationally, farm full-factor productivity rose by more than 2 percent a year during the war; productivity per hour of work rose even faster. Now, in a labor-scarce environment, local farmers had every reason to adopt all the latest tools and chemicals, and to help them do this, they could rely on years of agricultural extension and vocational education. As a student of vocational education, I tried to incorporate all the latest information on our farm and encouraged its use on several other local farms, some owned by relatives.

POSTWAR TRANSFORMATIONS

Prosperity during the war did little to assuage a deep anxiety felt by local farmers. What would happen when the war ended? Many remembered the price collapse in 1921. Everyone remembered the Great Depression and had scant hope of avoiding another one when postwar spending abated. Farmers did a good job of communicating these fears to their congressmen. In 1942, as a gesture to farmers and as a way to maintain or increase farm production, Congress guaranteed price supports for twenty farm commodities at 90 percent of parity for two years after the legal end of hostilities (this would be 1946) or until the 1948 crop year. Such a guarantee was not really necessary, for the devastated agriculture in much of Europe would take up to a decade to recover. In 1948 Congress extended the 90 percent guarantee until 1949 and voted for flexible supports after that year, with a 75 percent guarantee in "normal" years. When prices softened in 1949, for the first time since 1940, farmers became scared and pressured Congress for higher support levels. An ever-willing Congress gave in and maintained the 90 percent support level for another year. Then, with the increased demand caused by the Korean War, it continued this level of support until 1953 and then finally extended it to 1954. Thus, contrary to farmers' expectations, they faced no major decline in prices until a decade after the war. In fact, the 90 percent of parity actually increased farmers' incomes compared with those in most other industries. By 1954, the years of high demand relative to production were over, and from that time to about 2005 (with the exception of the mid-1970s), farm production outran demand, and problems of surpluses and support prices remained at the center of agriculture policy making.

In our village, the number of farms actually increased during and immediately after the war. This was contrary to national trends but reflected a local pattern. In most parts of the country, nineteenth-century farms typically included large woodlots and thus had room for expanded cultivation through the clearing of new ground. Although some farm consolidation took place, the usual pattern was a division of farmlands among heirs at the death of an owner. Thus, when Grandfather Conkin died in 1918, his will mandated the division of his large 150-acre farm into three roughly 50-acre farms, one each for his daughter and two sons. Just before the war, two large farms that adjoined ours went through similar divisions. The more developed farm, just up the road from our house, eventually ended up in six different parcels, none of which was large enough to provide a profitable farming income. On the other side, the only son of the owner kept the farmland intact, and until recently it supported a grade A dairy operation. But, typical of the postwar period, various members of that family took off-farm employment. The only other large adjoining farm—to the south—remained in the possession of a surviving daughter until after the war but was not intensively farmed. She then sold it in two parcels, one of which was large enough for full-time farming in the immediate postwar years. The owner of that parcel was one of the most skilled tobacco farmers in the area, but his family income was low in comparison to that of the few very large farmers or that of most local families in which the heads of household worked off-farm.

My family kept the farm operating near its prewar level for only a few years after the war. I grew tobacco through 1947, and my mother continued selling milk well into the 1950s. In most years we grew a small corn crop, and we had to put up hay for the cows. But most of our thirty acres of cultivable land was now in pasture. In the summer of 1947, just before my first year in college, I gained a job with the county office of the old Agricultural Adjustment Administration (AAA), by then named the Production and Marketing Association (PMA), and today called the Farm Service Agency. For four summers I measured (that is, surveyed with a chain) the tobacco acreage of slightly more than 300 allotment holders in our county district and, in most years, an average of 100 farms in adjoining districts. I was paid by the farm, and since most of the farms in our district were small, with allotments averaging just over an acre, I made good money. My sister also spent two summers working in the PMA office in Greeneville, mainly helping with the office work for the tobacco program. During my time with the PMA, I gained firsthand knowledge of the farming methods and problems in our district. I also became an expert in almost all aspects of burley tobacco culture.

In one sense, not much had changed in this district during the war, and this would be the case until well into the 1950s. Only two or three farms were larger than 200 acres, and none had a tobacco allotment of more than 3.5 acres. At least half had allotments under 1 acre, with a few as small as 0.2 acre. This was the result of the fragmentation of farms through inheritance or the sell-off by owners. In this period, the average price of burley tobacco was around 45 cents per pound. A superb farmer might grow 2,000 pounds on an acre, but the average was less than 1,500. Only a few rare farmers received as much as $900 for an acre (gross income, not net), which still required about forty-five days of labor. Even discounting the value of labor expended, the costs were considerable—fertilizer, insecticides, maintenance of tools. Yet no other crop was nearly as profitable on an acre-to-acre basis, and tobacco remained the principal money crop in our county. As late as 1948, the average net farm income was still under $1,000. Unskilled factory wages held at around 50 cents an hour during the war and rose only slowly afterward. My father earned an annual income of just over $1,000 at Eastman Kodak. Thus, on large farms with 3 acres of tobacco allotment, the net income from tobacco and possibly a small dairy operation could still compare favorably to urban jobs. But in my part of the county, almost no farmers had the land, tools, and allotments to do this well.

During my visits to farmers in this northern district of Greene County, I found only about fifteen or twenty really profitable farms. Up to half the farms were operated by men who, like my father, had full-time off-farm jobs. They all grew the allotted tobacco but increasingly shifted most of their cultivable land to pasture for dairy cows (the old pattern) or, increasingly, to beef cows (the wave of the future for almost all hobby farmers). With the small size of local farms, the only path to incomes above the national median was a combination of tobacco and an up-to-date dairy farm. This meant securing a permit to sell grade A milk from the local dairy association, which was not easy. Another possible path to success was new in our area. When I stopped growing tobacco in 1948, my father gave up farming altogether and rented the tobacco crop to a young man who lived two miles away, just over the Washington County line. He worked the land on shares, but this arrangement reflected the new pattern of sharecropping. He was already developing a large dairy operation (still the largest in our area) and rented available land on several scattered farms. He and a crew of workers came periodically, in trucks or on a tractor, to tend our tobacco patch. These workers were local young men who lacked a high school education and had few employment opportunities. Forty years later they would almost all be replaced by migrant workers, mostly Hispanics.

The fact that a young farmer rented land so far from his home base reflected the unique situation of tobacco farming. He was able to rent land close to his own farm for extra pasture or hay fields, but few small farmers would give up their tobacco patches. And until 1971, the burley tobacco allotment was tied to land. From 1934 through 1971, burley tobacco growers, largely in Kentucky and Tennessee, voted overwhelmingly in periodic elections to accept acreage controls and, with them, price supports. The allotment on each farm was based on acreage and past production records. This allotment was a valuable asset that increased the value of a farm (if sold, the allotment went with the land). In order to market a crop, one had to have a certificate from the county AAA (or its successors) attesting that the amount grown that year did not exceed the allotted amount. Legally, the elected county committeemen had to certify acreage, which meant doing the actual measurements. The fact that few of them chose to do so explains my job for four summers.

I visited each farm. The farmer, or an agent, had to be present and help move the chains used in the survey. I was not supposed to figure the acreage or tell the farmer how much he had. My job was simply to draw a diagram of the field, with all the measurements. In most cases, however, I could easily calculate the acreage and told the farmer what he wanted to know. What I failed to communicate to all but the most sophisticated farmers was that they would be better off if they planted a bit more than the allotted acreage. If the PMA office determined that a farmer was not within the limits of his allotment, it sent a letter stating the amount of the overage and sent a permanent employee of the county office to the farm to supervise the destruction of the excess crop. It cost a mere $3 to do this, and it allowed farmers to destroy the poorest parts of their fields, including areas drowned out by too much rain. This process ensured that farmers grew the full amount allowed. In all too many cases, farmers came in well under the allotment, and they were usually pleased by that fact, as if planting too much would have been a crime. I could not persuade most farmers to get a prior survey of a prospective field to ensure they planted the absolute maximum.

Along with the elite few, at least a hundred full-time farmers in our district still tried to make a living on the farm after World War II. Many of these were older men, moving toward retirement. A majority held on to the old ways. Few owned tractors before 1950. Those far off the main roads did not get electricity until after 1950. Tobacco and grade B milk provided most of their cash income. They still depended on home provenance. In most cases, the next generation would not remain on these farms. Those who inherited all or part of the land usually worked at outside jobs and moved from crop farming (with the possible exception of a tobacco patch) to hay and pasture. In other words, agriculture as a full-time occupation was doomed in hilly, northern Greene County. This fact became obvious only in hindsight and invited a degree of sadness.

But an important point is that the land remained. Almost all of it remained agricultural, in the sense that it produced goods for market and some income for owners, although in most cases the farm operated at a loss if one fairly assigned a value to the labor expended. In a few hilly areas, formerly cultivated land relapsed into young forests—a pattern that had prevailed in much of New England a century earlier. But this desertion of agriculture was not the rule. Indeed, up to half the young people left the farms and moved to nearby cities, where they found employment. But the other half, now within easy commuting distance of factory jobs, preferred a hybrid way of life. Ironically, the household income of families classified as farm operators by the Department of Agriculture steadily increased, even as the average net farm income declined. In fact, the household income of small or hobby farmers usually exceeded that of full-time farmers who continued to operate the best farms in the area. Because of its location, my village, only seventeen miles from highly industrialized Kingsport and almost the same distance from rapidly industrializing Greeneville and Johnson City, led the way in the most critical change in rural America—the gradual development of a single labor market embracing both urban and rural areas, accompanied by a complex array of lifestyle choices.

Home provenance gradually declined after the war—less so for the older, traditional farmers who kept their milk cows, chickens, hogs, large garden plots, orchards, and cellars full of canned vegetables and fruits, and more so for the few highly commercial and more specialized farmers and for part-time or hobby farmers. This shift reflected a desire for less arduous labor, higher incomes, and external shifts in the food system. I will use my family as an example of these shifts.

Through World War II, federal policy encouraged home sustenance, including victory gardens in urban backyards. Farm wives were given extra sugar rations for canning. Meat rationing supported homegrown pork, eggs, and broilers. But by the end of the war, the incentives for home production steadily lessened. Even during the war, my parents went to town almost weekly for shopping and took advantage of the lower food prices in emerging supermarkets versus the local country stores. Rapid advances in the processing, packaging, and transportation of foods gradually lowered the cost of fresh as well as canned or packaged foods. Home refrigerators allowed longer storage of fresh produce and meats. But the most important change was in the price of store-bought foods relative to income levels. This was a result of the increasing efficiency and productivity of commercial agriculture. Also important was the wider array of choices in terms of types of food and their off-season availability, as well as the health benefits of turning away from a diet overly dependent on pork and lard.

At the end of World War II, my mother stopped raising chickens. Soon most local families did the same. The revolution in poultry production was just beginning (a hundredfold increase in labor productivity in modern broiler factories). Thus, it was soon much easier to buy eggs and broilers than to raise them, not to mention being less expensive if one considered the labor expended. By 1950, nationally, the percentage of farms with chickens had declined from 95 percent in 1900 to 78 percent (today it is less than 1 percent). My mother continued to milk cows and sell grade B milk into the mid-1950s. Most full-time farmers also sold milk. But soon, almost no one kept a milk cow just for home consumption. Nationally, milk cows on farms had declined from more than 80 percent in 1900 to 68 percent in 1950 (today it is only 8 percent). Keeping a cow was simply not worth the time and effort, and people increasingly preferred homogenized and pasteurized milk, particularly for children. By the 1960s, fewer and fewer families continued the low-volume sale of grade B milk, and the number of dairy cows declined as small or hobby farmers switched to beef cows, as did my father in the late 1950s. At about the same time, my parents stopped raising hogs. By 1980, perhaps no more than a dozen local families still butchered hogs each fall or, more commonly, paid professional butcher shops to kill the hogs and process the meat for them. With this shift, meat consumption changed to a mixture of beef, chicken, and pork, all bought at the super-markets in Greeneville and Kingsport.

Locally, most families continued to grow vegetable gardens, but few maintained home orchards, except for some low-maintenance summer apple varieties (by 1992, only 92,000 farmers in the United States grew apples). My grandparents, who continued to care for a large home or-chard until they retired and sold their farm in 1955, had adapted to the new demands of successful orchard crops. They carefully pruned trees, grafted new varieties onto rootstock, and, most important, maintained a regular regimen of spraying to control insects and diseases. In earlier days, neither insects nor diseases had been much of a problem, in part because of older varieties, and in part because insects had been scarcer before the advent of new types of agricultural chemicals. Almost no other families in the area were willing to go to all the effort, and most of them bought their fruit from the few orchards clustered in the foothills of the nearby mountains or from supermarkets that increasingly sold fruit, off-season, from as far away as California.

Although vegetable gardens remained popular, food preservation quickly changed after the war. Two local developments secured this shift: the electrification of almost all homes, and the advent of frozen foods. The local TVA distributor had extended electric lines to all the homes in our village by 1950. This meant that everyone could have a refrigerator and a stand-alone freezer. The change came quickly. In 1945, as the war wound down, our high school agriculture classes began a major project to build a community cannery and, in the process, hone our skills in carpentry, bricklaying, plumbing, and wiring for electricity. We built a state-of-the-art facility with a large boiler, two large pressure-cooker vats, and tables and tools for food preparation. Our expectation was that farm-ers for miles around would bring fruit, vegetables, and even meat to the cannery, where, for a small price, they could buy metal cans like those used by food companies. The high steam pressure assured the safety of the contents of the filled cans. Unfortunately, use of the facility slackened after only two or three years. Frozen foods largely took the place of all the glass jars that had once filled root cellars and basements. For a few items, such as tomatoes and tomato juice, home canning remained popular, but not for long. Because home freezing was never as successful in retaining the flavor and structure of vegetables as the flash freezing of commercial food companies, the shift to store-bought frozen foods also increased each year.

Electricity reduced home labor in several other areas. It allowed the use of many household appliances, as well as pumps to bring well or spring water into homes. This meant that farm families who could afford it built bathrooms and retired the traditional outdoor privy. It also allowed the installation of electric heat. TVA electrical rates were among the lowest in the nation, and after the war, the local distributor launched a crusade to get farmers to heat with electricity. My family gladly followed this advice and, for the first time, could give up the long process of gathering firewood. The electric range also meant that housewives could prepare summer meals without the stifling heat of a woodstove and also enjoy the cooling effect of electric fans (it would be the 1980s before many local people bought air conditioners). Thus, in not much more than a decade, most local families had reduced the amount of work dedicated to home sustenance by at least 80 percent—few or no morning and afternoon chores (feeding, milking, gathering eggs, slopping hogs), no wood cutting for fuel, little canning or preserving of meats and vegetables. In 1935 farmers in the United States gained about 15 percent of their income from goods consumed at home (I would estimate that in my village this number was closer to 20 percent), and this percentage may have gone up during World War II. Today, it is less than 1 percent nationally. Although the gradual shift to beef cattle still required plenty of hay for winter feed and some daily chores in the coldest weather, the quick adoption of mobile hay balers after the war drastically reduced the work required to get hay into the barns.

From 1950 to 1970, American agriculture grew at an astonishing rate (described in the next chapter). These two decades marked the apex of a sectoral revolution. One result was the reduction of the agricultural labor force by about one-half. Another was the rapid consolidation of farms, with almost all the productivity growth taking place on larger farms. For the most part, the story of farming in my home community in east Tennessee is a story of those who were unable or unwilling to join in this revolution and thus moved to the margins of American agriculture. By 1970, only five farms out of the seventy or so that had been present in our village in 1940 came close to providing a median income for the farm operator, and each of these produced grade A milk. Only about ten other less productive farms continued with full-time operators until 1970. One large farm, owned by a Kingsport businessman, had a full-time manager during these years. Beyond that, it is hard to estimate how many other farms survived until 1970, because the answer depends on how one defines a farm. By lease or purchase, about forty former farms were effectively consolidated into larger productive units (most for hay and pasture). But at the same time, the continuing division of farms into smaller and smaller units meant that the number of people who owned farms (defined as at least ten acres with some annual sales) increased. Unlike in the past, ownership of farmland did not correlate closely with farm management or employment.

Government policy had a major role in the structure of local agriculture. This was true because of the continuing importance of burley tobacco. Acreage controls and price supports continued through 1970. Farmers could increase their tobacco acreage only by buying farmland with an established quota or when the local committees redistributed small amounts of acreage that had been dropped from tobacco production. As more and more farm owners worked off the farm in Kingsport or Greeneville, they either continued to grow their ever-shrinking tobacco patches or rented them to neighbors. They already had the needed infrastructure, such as a tobacco barn. By Department of Agriculture rules, the tobacco had to be grown on the farm that had the allotment, resulting in an absurdly fragmented crop, with the average allotment in our area dropping to less than one acre. Because of health reasons, the market for tobacco fell almost every year, and thus patches became smaller and smaller. Despite this shrinkage, the production never declined by the same proportion, since farmers had every possible incentive to grow more on less land (good soil, more fertilizer, more insecticides), and they succeeded. From 1950 to 1971, the average yield of burley tobacco rose from 1,350 to 2,400 pounds per acre, with maximum yields as high as 3,000 pounds. Thus, even as the amount of cultivated acres declined with each passing year in our community, almost every farm still produced tobacco.

This system came to an end in 1971, by a vote of burley tobacco growers. The new tobacco program involved marketing quotas based on pounds, not acres. More significant, these quotas could be sold to other growers in the county at whatever price one could negotiate. Since the offering price for poundage was only a small percentage (around 20 percent) of the value of the harvested crop, it was still in the best interest of some farmers to grow their own tobacco or rent it to neighbors. However, many farmers with tiny allotments stopped growing tobacco and gained what they could from the sale of their allotment. In some areas of the county, a few farmers bought enough poundage to plant ten or more acres of tobacco, using temporary laborers for the harvest. In other words, the new system allowed a gradual consolidation of tobacco growing on a few large, specialized farms. When Congress ended marketing

controls for tobacco in 2004 and compensated farmers who lost their allotments, almost all tobacco growing shifted to relatively few large farms.
In a final major structural change, the old auction system and the large
tobacco warehouses gave way to direct contracts between tobacco companies and growers, or a type of vertical integration that began to blur
the lines between growers and processors. Thus, tobacco's century-long
reign as the main cash crop in our community ended. It is rare today to
see even one tobacco field anywhere near our village.

With the shift away from tobacco, crop agriculture all but ended in
our area of the county, unless one counts hay as a crop. With beef cattle
dominating, grass hays such as fescue were sufficient for winter feeding.
Fescue is coarser and less nutritious but more hardy than timothy or orchard grass. Unlike legumes, which became less popular after the shift
away from dairying, fescue requires little care year after year, except for
fertilizer and lime. Even by 1970, despite the surviving tobacco patches
on the most fertile and usually noneroding land, the countryside was
now mainly green, and the runoff of flooding rains was almost clear
of silt. Environmentally, the community had improved dramatically. No
one cultivated the steeper hillsides. Old ditches recovered slowly from
past mistakes. But, on the negative side, the manure and urine from
grazing beef cattle, which waded in and drank from the valley creeks,
created enough water pollution to kill off most fish. Spring-fed streams,
whose water we often drank when I was a boy, were now full of deadly
bacteria.

Except for one dairy farm that borders our community, only two or
three families are fully dependent on farm income today. Almost everyone has off-farm employment of some type or is retired. The beef cows
present on every hillside may or may not belong to the owner of the
land. But the cow owners who rent the pasture are most often part-time
farmers. In fact, it is difficult to develop criteria that would enable one
to count the number of farms in our community. But it is important to
note that our community straddles a state highway, is well located for city
employment, and never had any large, well-situated farms. In a sense, it
has become a far-out bedroom community of Kingsport, where at least
half the local workers are employed.

State Highway 93 roughly parallels the creek and valley that make up
the heart of the Bethesda community. Those who attended the Bethesda
church and had a sense of belonging to the community lived on each
side of this highway for a distance of more than three miles. In the mid-
1930s approximately twenty homes were along this section of the road,

or about a fifth of the total number of homes in our community. The others lived on seven side roads, most of which were only recently paved, that fed into the main highway. Of those twenty families, sixteen were headed by full-time farmers. Today, there are more than seventy-five homes on both sides of the same stretch of highway. Of the fifty-five additional families, approximately forty-five live on lots of less than five acres. But at least eleven plots in excess of ten acres, some farmed by absentee owners, have split off from larger farms. Thus, by Department of Agriculture criteria, the number of farms may have actually increased along with this suburbanization. Even the number of farm operators may have increased (definitions are unclear here), even as the number of full-time operators has dropped to near zero. Admittedly, on the roads farther away from the highway, the increase in homes has been slower, but even there the decline in full-time farmers has been comparable.

This pattern was widespread in the upper South, particularly in Tennessee and North Carolina: an increase in landowners, a decrease in full-time or profitable farming, and strip housing developments along major roads. What one sees as one flies over the former farming areas of the Piedmont or in my area of east Tennessee is a complex urban-rural mix, or what I call a "rurban" pattern. After World War II, this area became the largest manufacturing belt in the country. But the manufacturing was located mainly in small towns, with workers driving in from what had once been open countryside but was now experiencing a steady growth in population. By choice, many former farmers chose to remain on the land and keep a hand in agriculture, if only to grow a large garden and pasture a dozen beef cattle. This was a lifestyle choice, and one that was not necessarily economically rational. They might sell a few lots along the road but keep the back acres in pasture. Many city people yearned for acres in the country, including professionals, and they gladly farmed their land at a loss each year (and received a tax deduction). For them, farming was a hobby. But full-time farming was not an economic option for most people in our village. To earn a median income from farming required an increasing amount of land, and because of its location near manufacturing and service jobs, that land was becoming more and more expensive. The capital investment required was simply too high, except for more remote land or for prime agricultural land, which was scarce in our area of Greene County.

The gradual transition from full-time to part-time farming in my home village was not a story of economic stress or failure or an example of rural decline. Because of employment opportunities in nearby towns, the household income of almost all the remaining part-time farmers rose with each passing decade. The farms contributed little to this income but clearly provided nonmonetary rewards. I am sure that most present residents, who did not live through the transition, suffer no sense of loss. But their complacency should not conceal what was indeed lost for those who, like my father, had to accept factory employment, not willingly but because of economic necessity. The key word is *employment*. The main concern was status and self-identity. What my father always lamented was a loss of independence, of an entrepreneurial status. In his own words, he was no longer his own boss, able to control the timing and tempo of work. He continued to tend his beef cattle as a way to retain some sense of a past proprietary role and looked forward to retirement, when he could at least play at being a full-time farmer.

Successful Farming in Pennsylvania

My sister would end up on a profitable farm not in Greene County or even in Tennessee, but in York County, Pennsylvania. Her story illustrates the requirements and the costs of competing successfully in the new agriculture. A high school classmate of mine, John W. Hunt, moved to Pennsylvania with his parents in 1947. He would later marry my sister Lois, his high school sweetheart. The reasons for the Hunts' move from their farm in Washington County, Tennessee, were complex, but among them was the inexpensive land in York County just after the war. John's father had worked on a family farm but had not owned it. He supplemented his farm income by running a milk truck. He followed other families from our area who had already moved to Pennsylvania Dutch country, actually buying a 260-acre farm from one such family. Of course, he had to go into debt to make the purchase, but in a few years, rapidly rising land values had increased the value of his farm by at least 50 percent. He was lucky in terms of the timing. I first visited the Hunt family, and their farm, in the summer of 1950. Here is what I observed.

The farmhouse and a large part of the land were in the borough of Seven Valleys, about eight miles southwest of York, with some of the land stretching into two other townships. The land straddled the south branch of Codorus Creek, which was large enough to qualify as a small river by most definitions. More than ninety acres of the land was in the floodplain of this creek; half of it too low lying for cultivation, but it made good pasture. Some of the sloping meadows were cultivable and assured good yields of corn and small grains. A large, hilly woodlot was to the north;

to the south, across the road, were more gently sloping fields suitable for all types of crops. The two-story, ten-room brick house dated from before the Civil War but, because of its age, required a good bit of maintenance. It was on a main road, or what would later become a state highway. The classic Pennsylvania Dutch barn was equally ancient and was one of the largest in the area. It measured 40 by 100 feet and had a rock wall base surrounding all the stables, with the back wall fully underground. Huge beams, some more than 40 feet long, supported the main barn above, with two large doors at ground level. The high roof, the shingled walls, and the numerous vents and windows allowed the storage of enough hay for 100 cows, plus all types of farm machinery. A silo at one corner held more feed. A large, slatted corncrib was close by. A small creek ran through the barn lot. An annex to the barn consisted of a milk shed, with stanchions for about 20 cows and a small hayloft overhead.

The Hunt family was thus blessed with what promised to be a profitable farm, one that was larger than most of the neighboring farms owned by older German families. The Hunts established a mixed farming operation. They briefly experimented with truck crops—peas and tomatoes—for a nearby cannery but found that the hand labor required for tomatoes was harder than that for tobacco back in Tennessee. From the beginning they also grew corn, wheat, and hay, and for a few years they kept hogs both for home use and for sale. They sold part of their corn crop and all of their wheat, and they baled and sold straw. But the main cash crop, as it was for most of their neighbors, was milk from a herd of twenty or so cows. John, the older son, soon established a milk route to supplement the farm income. The farm, so close to York, had electricity but lacked indoor plumbing. It was, in this respect, comparable to the farm the family had left in Tennessee. When they moved in 1947 they brought their Ford tractor with them and would soon buy additional and updated harvesting equipment. They used portable electric milkers and still strained the milk into the standard ten-gallon cans, but they had chilled water for the cooling.

In the 1950s, as the agricultural revolution took off, the Hunts continued to operate what became primarily a dairy farm. They had to upgrade operations to meet higher sanitary standards set by dairy associations in both the Philadelphia and New York districts. This meant an improved milk house, with a stainless-steel tank for cooling, and rigid rules for cleaning the milkers and caring for the cows. In 1956 John was drafted into the army; he had received a deferment during the Korean War because of his vital work on the farm. In 1957 John's father died suddenly of a heart attack at a relatively young age, and John was able to get an early discharge and come home to take over management of the farm. To add to his burdens, his mother died of breast cancer not long after. It was at this point that he resumed his earlier courtship of my sister, who was then secretary to the vice president for patents of Tennessee Eastman in Kingsport. Lois had graduated from East Tennessee Teachers College with a joint degree in home economics and business. They married in early 1959 and settled down on the farm in Pennsylvania. John had already made the difficult decision to go into debt and buy the two-thirds of the estate owned by his brother and sister. Thus, the couple made a commitment to a career in an increasingly specialized and capitalized agriculture in what had long been a very healthy farm economy, particularly in Lancaster and York counties. John would remain a full-time farmer until his early death, from cancer, in 1994. For more than a decade my sister was, in a sense, the business manager of the farm, keeping all the records and ensuring that the farm took advantage of all the benefits offered by the federal government, as well as being a homemaker and eventually the mother of four children.

Despite all the advantages of a large, well-located, and fertile farm, it was still difficult to earn the income needed for a large family. The only answer was to grow by buying more land, purchasing the latest and best machinery, and, above all, increasing the dairy herd—the largest source of income. In 1964 John and Lois had an opportunity that was too good to pass up. The owner of an adjoining dairy farm decided to sell his land, in part because he was of retirement age and his farm was too small to provide a good living. It joined most of the southern boundary of the Hunt farm and had been a part of their home farm in the years before the Civil War. Land values were already rising at a rapid pace, so investing in the farm seemed likely to be profitable, but the immediate problem was how to pay for it. John and Lois folded their older mortgage into a new one and began making regular payments. The new farm gave them a second house and an older but quite useful barn. They now had almost 400 acres of very valuable land.

Above all, the new land enabled John to expand his dairy farm to around seventy cows, considered a midsized dairy (locally, a large dairy). The additional land was mainly on sloping hillsides, with little meadow or pasture. But the contoured fields, some with terraces funded largely by federal grants, approximately doubled the amount of cropland. John also built a new milk barn at the same time he bought the neighboring farm. The new barn had stanchions for forty cows (John later regretted not opting for a milking parlor instead) and a pipe system that took the milk directly from the milkers to a new and larger stainless-steel cooling tank. It had a reverse system that, after milking, automatically cleansed the lines. The workload far exceeded the capabilities of one person, but the second house enabled John to employ and provide housing for a full-time assistant, who did half the milking for a share of the gross income from the milk.

By 1974, the Hunt dairy farm was the largest taxable property in the borough of Seven Valleys. But even 400 acres was not enough land. It did not fully utilize and thus justify the expensive equipment, which included a large truck, three or four tractors, a chopper for silage, elevators, manure spreaders, several wagons, special mowing machines that crushed the stems of hay, special balers that tossed the bales into a wagon attached behind, various plows and disks, a special corn planter for no-till corn, and, most expensive, a combine with heads for both corn and small grains. The transition in agriculture, the mandate to grow or die, led many nearby farmers to give up. Some sold out, but many sold only some land and earned extra money by leasing what was left to farmers like John who had the equipment to cultivate it, usually for corn or possibly soybeans. Thus, by the 1980s, in some years John was leasing as many acres for corn as he cultivated at home. He also did custom work for smaller farmers, particularly combining. But in a rapidly industrializing and growing York County, land values began to soar. Thus, land along major roads, with huge developmental possibilities, became so valuable and the taxes so high that farming income provided nothing even close to a competitive return on the committed capital. Fortunately, the state and county maintained a special, lower assessment value for agricultural land, which allowed some farmers to survive. But by the time John died, he could have sold his farm and equipment, invested the money conservatively, and still increased his annual income. Such was the fate of agriculture in almost all metropolitan areas (in counties that were part of a statistical metropolitan area).

In 1974, at the peak of his dairy operation, John decided to give it up and sold his cows. They had provided him with a reliable income, but the work—including twice-a-day milking—was very demanding. He was having trouble with his knees from years of kneeling to wash the udders of his Holsteins (a milking parlor with a pit might have prevented this). But even with 400 acres, it was difficult to earn a living from other types of farming. Thus, my sister resumed her office work, eventually becoming secretary to the president of the American Chain Company in York. Her off-farm employment became an important supplement to the family income. In some years John bought young beef cattle and fed them off for market. This feeding operation could be profitable in one cycle, a loser in the next. John also intensively cultivated his land, rotating fields among corn, winter wheat, soybeans, and hay (crop rotation remained the norm in the hilly fields of York County). He cultivated up to 200 acres of corn at home and in rented fields. Like most farmers, in some years he hedged his returns by selling the corn before he even grew it, sometimes gaining on the futures market. He continued to use the old-fashioned rectangular bales for both hay and straw rather than the new huge, rolled bales. The extra labor paid off because of the large market for timothy hay and bedding straw among the many horse farms in southeastern Pennsylvania. In some years, straw sold for as much as hay and brought in as much income as the wheat itself. John's income was not large, but with Lois's salary to count on, the family lived well. His was among the largest local farming operations and was more profitable than most of the Amish and Mennonite farms just east of his. Meanwhile, the 400 acres increased in value each year, so his wealth grew much more rapidly than did his income. Also growing were the trees in his extended woodlots, with the value of timber being an often overlooked asset.

In almost all respects, my sister and brother-in-law were in a position to maximize the opportunities in one of the first great agricultural areas in America. John had studied agriculture in high school, avidly read a bundle of farm journals, and effectively adapted the newest tools and techniques, including the multiple chemicals (particularly herbicides) now required in agriculture. My sister was an excellent financial manager; she reported all their income to the IRS but also took advantage of all the tax breaks offered to farmers (such as shortened depreciation schedules) and all the government payments they were entitled to. Yet, even with all these advantages, their annual net income from farming (not including capital appreciation on the land) was well below the median income for that area of Pennsylvania. Only Lois's off-farm income pushed their household income above that median. Given these realities, it is understandable that every year, somewhere in the Seven Valleys area, John attended the farm auctions held when traditional dairy farmers with less land and fewer skills had to give up. The squeeze was on, and John's survival was a notable achievement. Today, two of his sons continue to farm, concentrating on corn, wheat, hay, and beef cows. Both also have full-time jobs in the York area.

At the time John gave up his dairy operation, a revolution was under way in dairy farming—one that he was aware of and troubled by. He considered the possibilities of what most called "confinement" dairying, which involved keeping the cows in barns or on small lots, off pasture, and bringing all their food to them, including grass, silage, and ground feed. With this system, one could buy all the necessary feed and keep large dairy herds on small plots of land near large cities, as was happening in California. By the time of John's death in 1994, his type of dairy operation was already becoming a bit of an anachronism. Yet in his area of Pennsylvania, it has survived.

According to the Farm Census of 2002, more than 25,000 of the 92,000 dairy farms in the United States had between 50 and 100 cows, or the size of the Hunt farm. The average acreage of these now "small" farms ranged from 260 to 499 acres, and the median value of the milk sold was well over $100,000 a year. But these small dairy operations contained less than 20 percent of all dairy cows and an even smaller share of total milk production. I believe that many of these formerly large, but now comparatively small, dairy farms will survive. They are the backbone of the farm economy in many counties, particularly in Vermont, Wisconsin, and Minnesota. They have strong local support. Concern for such small operations is also present in Congress, as reflected in a new program in the 2002 farm bill—special market loss payments to dairy farms with fewer than approximately 132 cows when grade A milk prices fall below the established Boston class I level. But one qualification is in order: such dairy farms simply cannot provide the income needed by a family that aspires to anything close to the average per capita consumption that is now standard in the United States. Thus, such farms will coexist with nonfarm employment by at least some family members.

Today, the degree of concentration and the scale of dairy farming are moving toward that which dominates in the raising of poultry and hogs. In 2002, 2,902 dairy farms had more than 500 cows, and almost all had annual sales of more than $1 million. The average herd size for farms with more than $1 million in sales was 1,500 cows. In total, these farms accounted for more than 45 percent of all milk cows in the United States. This trend toward the consolidation of farms and increase in scale is happening all over the United States in almost all areas of farming. In the next chapter, I explain not only what has happened but also some of the necessary conditions for an unprecedented agricultural revolution.

5. Dimensions of an Agricultural Revolution

If one defines an industrial revolution as a dramatic increase (at least 50 percent) in full-sector productivity during a single generation, then American agriculture has attained the most important industrial revolution in American history. Just like manufacturing or services, agriculture includes many distinct industries (or crops). Shifts in production have not always been synchronized among these industries, with some crops lagging behind others. For example, cotton, and with it the South as a whole, lagged behind the rest of the United States, at least until the 1950s. But after 1950, very few crops failed to experience an almost unbelievable burst of productivity gains, with the largest gain in almost every case in production per unit of labor (less so, but still large, for inputs of land and capital). Dates are a bit arbitrary, but in no other twenty-year period were the agricultural gains quite so dramatic as they were from 1950 to 1970. Change was so rapid that almost no one was able to measure, or comprehend, what was happening.

By most estimates, some based on thin evidence in the nineteenth century, the annual growth in full-factor productivity in agriculture rose, on average, about 1 percent a year in the century before 1935. All economists agree that, in about 1935, but clearly by 1940, this annual growth rate at least doubled and remained this high until 2000. And all economists agree that labor productivity was more rapid than returns on land or capital. What economists have not agreed on is the exact rate of this accelerated full-factor growth. For a heterogeneous agriculture, a definitive answer may be impossible. The late Bruce L. Gardner clarified many of the problems involved in making such an estimate and concluded that the full-factor rate was around 2 percent. Economist Sally H. Clark, in the most robust estimate of full-factor growth, places it at 3 percent from 1935 until the mid-1980s. This growth was more rapid than in any other economic sector and more rapid than that in all but two of the twenty largest manufacturing industries. She believes that the rate of growth for labor productivity was a hefty 4.5 percent, a truly revolutionary surge unrivaled by any other sector or major industry. What is amazing is the long-term endurance of this growth. Since 1950, labor productivity per hour of work in the nonfarm sectors has increased 2.5-fold; in agriculture, 7-fold. In one generation, from 1950 to 1970, the workforce in agriculture declined by roughly half, while the value of the total product increased by approximately 40 percent.[1]

Even these numbers understate the pace of change. With each passing year, the output of small farmers (most with off-farm employment), who constituted more than half the 2.9 million farmers in 1970, produced an ever smaller proportion of the total output. This trend continued, so that by 2002, only 322,625 farms produced 89 percent of all output (this is based on the value of what was produced, not on the pounds, bushels, or tons). But it is important to note that at least 425,000 principal or secondary operators managed these farms and that such large farms employed almost two-thirds of all farm laborers, most on a part-time basis.[2]

The gain in efficiency, based on labor input, was almost unbelievable in some crops. In 1900 it took 147 hours of human labor to grow 100 bushels of wheat. By 1950 this had shrunk to only 14, and by 1990 to only 6. For corn, the number of hours per 100 bushels shrank from 147 hours in 1900 to 16 in 1950 and to 3 in 1990; for cotton, from 248 hours per bale in 1900 to 100 in 1950 (just as the new mechanical cotton picker began spreading to more and more farms) and to only 5 in 1990. The labor required to harvest 10 tons of hay fell from 65 hours in 1950 to only 11 in 1990. In some livestock operations, the increase in efficiency was even greater, with chickens leading the way. In 1929 it took 85 hours of work to produce 1,000 pounds of broilers; by 1980 it took less than 1 hour. Comparable gains are now taking place for hogs. Based on Department of Agriculture estimates, the output per worker on American farms grew by 68 percent in the 1950s and by 82 percent in the 1960s. Output per farm rose tenfold from 1935 to 1997 for field crops (note that this reflects farm consolidation, and thus fewer farms with each passing year), and the rate was much faster for poultry and hogs. Such gains could not be duplicated for all crops, including tobacco and many fruits and vegetables.[3]

Almost as dramatic was the surge of the yield per acre for cultivated crops, which was a necessity for the productivity gains achieved since 1950. The yield for corn, our largest national crop, rose from around 25 bushels per acre in 1900 to 40 bushels by 1950, with the impact of hybridization; it doubled to 80 bushels by 1970, with the dramatic effect of herbicides, and exceeded 120 bushels an acre by 2000. The wheat yield rose from about 19 bushels in 1950 to 36 bushels in 1970. Cotton, which yielded around 270 pounds per acre in the five years before 1950, nearly doubled to 513 by 1972. From 1950 to 1970, annual milk production per cow almost doubled.

The size of farms gradually rose, although in many cases the total acreage reflected both owned and rented acres. In 2002 the average size of each farm was 441 acres, with the average value of what it produced at $97,000 (a number sharply reduced by small farms). In only one major area—poultry—does farm size not correlate at all with productivity and income, although the same phenomenon is increasingly true for hogs and dairy cows. For example, in 2002, 2,453 farms of less than 10 acres had annual poultry sales of more than $50,000, but only one farm of similar size had comparable income from corn. In 2002 just fewer than 100,000 farms had as much as 1,000 acres; 19,000 had more than 2,000 acres. These very large farms produced 40 percent of the total agricultural product.[4]

How does one account for such a surge in productivity and scale? Most economists believe that past investment in research, development, and education were indirect causes. Human capital may have been all important. But given at least a minority of highly skilled farmers, I believe that rapid changes in four areas provided the necessary conditions for this long-term trend in growth: machinery, electrification, chemical inputs, and plant and animal breeding. By necessary conditions, I simply mean that without one of these four conditions the pace of growth would have been much slower. Governmental agricultural policies had to provide an institutional framework for this growth. How much such policies abetted, or possibly inhibited, this growth is the topic of the next chapter. It is important to note that many of the factors supporting productivity growth in the United States were soon present in most parts of the world. The agricultural revolution was a global phenomenon, although nowhere else was the pace of change so rapid.

THE GREAT NEW MACHINES

New or improved tools were essential elements of growth. After World War II and the devastation of Europe, American companies dominated the market for farm implements. The competition was intense, and the pace of innovation unprecedented. The result was a level of technological sophistication undreamed of in the past. Without such technological prowess, neither the United States nor the world as a whole could feed the 6.5 billion people now on the earth. Hundreds of new machines increased labor output in every crop. American farms soon became so capital intensive that the cost of entry into farming discouraged most aspirants. What provided the energy to run all the machines was the burning of hydrocarbons, most from fossil fuels. The new tools required more specialized skills from farm operators, exponentially increased the amount of land needed for efficient farms, and widened the gap between highly efficient and specialized farmers and those who could not compete. Farming, at the highest level of output, is arguably our most demanding profession.

In the immediate pre–World War II era, the most important technological innovation clearly was the tractor. But only after 1950 did tractors, or self-propelled harvesting machines, fully displace horses. By then, the modern tractor had matured. Only refinements and added size and power have marked the last fifty years. By 1945, either bankruptcies or mergers had reduced the number of farm tractor producers to only seven domestic companies: International Harvester, John Deere, Case, Ford, Allis-Chalmers, Oliver, and Minneapolis Moline. A Canadian company, Massey-Harris, offered strong competition, while Caterpillar dominated the market for crawler-type tractors. Subsequent mergers have reduced this number to only four domestic companies, but they are now challenged by a much larger number of foreign brands.

Today, tractors, many with eight wheels and four-wheel drive, provide almost all the energy required for soil preparation and planting of all major crops, for all haying operations, for almost all types of hauling, and, until herbicides eliminated most of this work, for the cultivation of crops. The number of tractors increased from around 3 million in 1950 to a peak of 4.7 million in 1970; since then, the expansion has been in tractors' size and power, not their number. Tractors, particularly on small farms, still pull and power harvesting equipment, but on large farms they have largely given way to self-propelled harvesters. And it was in the harvesting of crops that the most important postwar innovations occurred.

Just after the war, pick-up balers and barn elevators cut in half the work of haying, while after 1970, the use of huge, round bales allowed one person to harvest a hay crop, since one could safely leave these bales out in the weather. Various vegetable harvesting machines, such as those used for lettuce and spinach, all but merge harvesting and processing.

Some of these huge machines can clean the vegetables and even pack them into boxes in the field. Although some crops, such as burley tobacco and certain vegetables, still defy machine harvesting, in almost all cases harvesting is now facilitated by machines, resulting in an enormous reduction in human labor and a huge shift from labor to capital inputs. Some machines harvest root crops, such as potatoes and sugar beets. Cherry pickers speed the harvest of fruits, while tree shakers help harvest nut crops. A type of small combine even harvests tender vegetables such as snap beans or peas or captures cranberries from their flooded bogs. Mechanical tomato harvesters pick most tomatoes (these are usually green and not very flavorful) and depend on new varieties that set tomatoes at one time. In almost all cases, machine harvesting has all but mandated a shift in scale toward larger and larger operations. It did not lead to a shift away from family-owned farms but to farms that, among major producers of field crops, routinely average more than 1,000 acres.

By far, the two most significant harvesting machines are the combine and the mechanical cotton picker. Both had their greatest impact in the period from 1950 to 1970. The combine was an old tool by 1950 (see chapter 1). It had revolutionized wheat production in the West and on the Great Plains, where it perfectly fit the short-stem grain that was usually very dry when mature. By World War II, a few combines were used on the farms of the eastern United States, but they were usually smaller than their western counterparts. In the humid East it was not easy to get tall wheat or other small grains ripe and dry enough for the combine without the risk of falling and even rotting. After the war, new drying equipment in grain elevators helped relieve this problem, with farmers deducting the cost of drying from the price they received for the grain. Also, smaller eastern farmers wished to retain their straw, which was not easy with combines until the perfection of pick-up balers after World War II. Thus, it was only after 1945 that the combine replaced the binder and the threshing machine in all grain-producing regions. Until then, most combines had been pulled by tractors, as some still are on small farms. But even the smallest of these machines were very expensive and not cost-effective on small farms. This left two alternatives for grain farmers—increase the scale of production by buying or renting more land, or hire custom operators to do the combining. Both occurred.

Two innovations dramatically increased the role of combines. One was the development and marketing of self-propelled combines. The first self-propelled combines were used in Argentina and Australia as early as the mid-1920s. Just before World War II, a Canadian company, Massey-

Figure 8. Contemporary John Deere combine with small-grain head. (John Deere)

Harris, developed a new self-propelled combine. It shipped 500 to the
Great Plains in 1944 and, with great publicity, formed a harvest brigade
to do custom combining from Texas to Saskatchewan. By the end of the
war, almost all manufacturers introduced self-propelled versions that
were larger and more efficient than tractor-drawn models. Instead of a
trailing cutter bar, which had to be offset from the tractor (otherwise, the
tractor would trample the grain), the cutter bar was now centered on the
front of self-propelled combines, allowing grain heads as wide as thirty-
six feet (see figure 8).

The second major development was the introduction of combine
heads that would strip and shell corn. The self-propelled combine was
perfectly adapted for this purpose, but the technological requirements
were intimidating. Like for small grains, elevators had to develop ways to
dry shelled corn, for in some years it was almost impossible to have fully
dry corn by harvest time. John Deere first produced corn heads for one of
its self-propelled combines in 1953 and began marketing them the next
year. Over the next five years these were perfected and became increasingly common on large farms (the heads were very expensive). By 1965,
combines shelled more than half of all corn; by 1970, they shelled almost
all of it in the Corn Belt. Today, the largest combines have heads for up to
twelve rows and can shell more than 500 bushels an hour (see figure 9).

Figure 9. John Deere combine with corn head. (John Deere)

They replaced the earlier machines that only harvested corn ears, leaving shelling as an extra step. With the perfected combine, all the major grains at the time of harvest went into large pools held in elevators. No one owned any particular grain; farmers claimed only a certain number of bushels in the pool. Specialized combines can harvest almost all seed crops. By changing screens to fit the size of the seeds, they can harvest most varieties of grass seed, soybeans, sorghum, and sunflowers.[5]

Given the versatility of combines, they are clearly the supreme implements in contemporary agriculture. No other machine, with the possible exception of the mechanical cotton picker, has displaced so many farmworkers, so lowered the unit cost of crops, and so directly affected the structure of farming in America. Their level of technological sophistication is almost beyond belief. For example, recent self-propelled combines, with their luxurious, air-conditioned cabins and airplane-like array of gauges, can provide a detailed nutrient map of every field harvested. Sensors record the minute-by-minute flow of grain into the bin, while a global positioning system (GPS) records the exact coordinates that match the flow data. This allows a farmer to so program his drills as to increase the flow of fertilizer to lean areas during the next planting season.

The largest combines cost more than $250,000. No small farmers can afford such a machine unless they do extensive custom work on dozens of neighboring farms. Even small, tractor-pulled combines, which are still on the market, are too expensive for the limited time they are needed on a 300-acre farm. Thus, in the eastern United States, the usual pattern is for a few large farmers to buy combines and use them for custom combining on nearby small farms. In the major Wheat Belt of the Great Plains, the dominant pattern is custom combining, or a throwback to the pattern of the old threshing machine. Gangs of large, self-propelled combines follow the wheat harvest from Texas to North Dakota each summer. A wheat farmer does not need to have any direct contact with his land. The owners of custom machinery can, in one operation, prepare the soil and drill the wheat. Thus a wheat farmer with 10,000 acres in Nebraska might live in Omaha and never even see his wheat fields. In this case, growing wheat is comparable to a high-risk investment, but the risk is considerably reduced by various government programs.

Another harvesting machine, though not so widespread in use, had an even greater regional impact than the combine. A machine that could harvest cotton revolutionized the American South and slowly restructured the cotton-growing industry. After a hundred years of experimentation, a few inventors had developed nearly successful machine pickers before the Great Depression. These first machines, available by the 1930s, stripped the bolls from the plants and then separated out the lint. Such "strippers" were well adapted only for the short, very dry cotton plants of Texas and Oklahoma, where some are still used. Elsewhere, farmers needed machines that could pull the lint from the bolls, or what would be known as spindle machines. In Memphis, John and Daniel Rust had working spindle machines by 1936 and demonstrated them on nearby Mississippi farms in 1942. Their Old Red picker of 1943 is now exhibited in the Smithsonian (see figure 10). In 1944 one of their machines harvested all the cotton on one large farm. Wartime restrictions allowed only limited commercial manufacture, with International Harvester selling about 1,000 machines during the war. These machines, which used wetted spindles and, later, burred spindles to pull the lint from the cotton boll, attracted both bountiful publicity and fear. It seemed as if the whole plantation system might be at risk, and both owners and sharecroppers were concerned. International Harvester began a large commercial production of one-row pickers in a plant near Memphis in 1947, while John Deere introduced a two-row picker in 1950.

In 1950 the new machines began gradually to replace the long, tedious hand labor involved in the cotton harvest. The first machines had some critical limitations. They collected so much trash (up to 250 pounds

Figure 10. Old Red, a 1943 International Harvester cotton picker, one of the first successful models. (National Museum of American History, a gift of Producers Cotton Oil Company)

per bale) that few existing gins could accept the dirty bales. Also, the early machines left at least 15 percent of the cotton in the fields. Thus, wholesale adoption of the new pickers required several parallel improvements—more efficient pickers, new gins with more elaborate cleaning devices, pre- and postemergent herbicides to reduce weeds, defoliants to strip the plants of leaves just before harvest, and new varieties of cotton that were taller and matured more uniformly. It is not surprising that the new machine had its first major impact in California, where irrigated cotton was a new crop. Half the California cotton was harvested by machine as early as 1951. Nationally, only 10 percent of cotton was harvested by machine as late as 1957, but almost half by 1960. The greatest lag in use was on the smaller farms of the Southeast, where the crop was already in steep decline. But by 1971 the revolution was just about over, with more than 90 percent of cotton picked by machine. For no other crop were the two decades from 1950 to 1970 so critical.[6]

Cotton pickers experienced the same shift in size and power as did combines. Self-propelled from the beginning, the machines grew in size with each decade. Today, some machines can pick six or eight rows, and a few, adapted for close-row planting, can harvest twelve. The cabs of pickers resemble those of combines, and GPS technology is beginning to allow the same type of field mapping. Complementing the picker is a new technique of getting the cotton to gins. More than 90 percent of cotton is now emptied from the bins of pickers into long, trucklike containers called modules, in which special tamping devices compact the cotton into tight, weather-resistant, truck-size bales that can be left in the field for days before tractor trailers haul them to the gin.

Structurally, the picker helped concentrate the Cotton Belt in the more expansive farms of the Mississippi Delta, the plains of central Texas, and the irrigated fields of the San Joaquin Valley of California. It was never useful on small cotton farms and in small fields. Thus small farmers either stopped growing cotton or sold out to larger farmers. In the Piedmont, few cotton farms survived except in parts of Georgia. In the 1930s around a million farmers grew cotton. By 1950, just as mechanical pickers began to have a wide impact, the number of cotton farms was down to just over 300,000. By 1974, when almost all cotton was picked by machine, the number was 80,000. In 2002 the number of cotton farms had shrunk to fewer than 25,000, with only 11,000 large farms producing 85 percent of the total. In much of the South, soybeans have all but replaced cotton, which has faced intense foreign competition and suffered from the use of synthetic fibers.[7]

Gradually, the mechanized cotton industry eroded the old sharecropping system of the South, even as it drove small farmers out of business. New Deal farm policies and wartime labor scarcities in the North had already lured more than 3 million migrants from the South to the North between 1930 and 1950, before the cotton picker had much impact. Approximately 2 million of these migrants were black, and almost all were from rural areas. From 1950 to 1970, in the midst of the technological revolution in cotton, more than 4 million southerners migrated, half of them black. This was the greatest internal migration in American history. These migrants in the 1950–1970 period often left not because they were pushed out by farm owners but because of the pull of northern jobs and, for blacks, greater equality. The push and pull were always interactive, but the end result was a transformed and much more efficient southern agriculture and a rapid reduction of farm labor in the South, with almost all blacks leaving agriculture (only 30,000 black farm operators today). Even among historians, the images associated with this migration vary immensely. On one hand, there are those, like Pete Daniel, who portray the migrants as victims of a modern enclosure movement facilitated by federal agricultural policies; on the other hand are those who rejoice at the liberation of former small farmers and sharecroppers from poverty, drudgery, and, for blacks, discrimination. Either way, in no other crop has the impact of new agricultural technology been so powerful, or the effects on American society so dramatic. Since 1950, the percentage gains in farm production have been greatest in the South, which has steadily narrowed the formerly large gap in production and income. Thus the rapid convergence of southern agriculture with national norms after 1950 was a major aspect of an industrial revolution, with all its benefits and all its human costs.[8]

ELECTRIFICATION

The second major requirement for an agricultural revolution was electrification. Only 11 percent of farms had electricity in 1935. The rural electrification program helped raise this to around 35 percent by World War II. The subsidized expansion after the war was very rapid, with 85 percent of farms electrified by 1950 and 97 percent by 1960. Rural telephones, which lagged behind electricity, gained federal subsidies in 1949 and, by 1964, were present on 76 percent of farms. For the major field crops, electricity might not seem very important, except in the critical area of irrigation. Electric motors replaced the windmill, powering the pumps for deep wells and providing the water pressure for huge spray-type irrigation. In the barn, electricity did far more than provide lighting; it powered elevators for hay bales or for silage, grinders for sharpening tools, air compressors, and many different shop tools.

It was among livestock farmers that electricity proved most revolutionary. The greatest productivity gains in modern agricultural history would be inconceivable without electricity to warm and cool the caged animals, power the automated provision of food and water, and provide lighting in what amount to poultry and hog factories. Electricity is also at the heart of modern dairy farming. It pumps the water, powers the suction milkers, cools the stainless-steel tanks, ventilates the holding barns and milking parlors, heats the water for the sanitary cleansing of lines and milkers, pulls the manure out of holding barns, and automatically dispenses feed (as it does in large beef-feeding operations). In the farmhouse, which has always been part of the total production enterprise, electrical appliances reduced the work of wives or servants, helped displace many domestic workers, and, through the radio, television, and now indispensable computer, provided the necessary market information for farmers. Record keeping is a vital part of farm operations, as is careful tax planning, all facilitated by a well-equipped farm office.

Chemical Inputs

New chemicals provided the third necessary condition for an agricultural revolution. This is a complex subject and involves five categories of chemicals: fertilizers, insecticides, fungicides, herbicides, and medications. Of the five, fertilizers and herbicides had the greatest impact on the production of crops, with medications presently having the most profound, but often highly controversial, impact on livestock.

Fertilizers

Plants depend on soil nutrients. Some new soils have an abundance of nutrients based on their original chemical composition supplemented by the past decay of plant materials. But sooner or later, heavy cropping exhausts most of these stored nutrients, and soil ceases to be productive. Even ancient farmers understood this reality and worked out methods of coping with it. One was a rotation pattern that allowed land to lay fallow (at rest from soil-depleting, heavy-foliage crops) for one or more crop seasons. This allowed wild vegetation to grow, decay, and add organic matter and nutrients to the soil. Equally important, farmers spread manure on fields to add nutrients, and they sometimes plowed under lush vegetation as a type of green manure. They buried seaweed or fish in hills to nourish crops and learned the value of ground-up bones. By the eighteenth century they realized that legumes enriched the soil (they store nitrates in airtight nodules on roots) and grew clover for that purpose. They also knew that the application of marl (a type of very alkaline soil) to some fields could increase the yield. It made the soil "sweet," as they said, which meant that the calcium carbonate in the marl reduced soil acidity. Today, lime serves this purpose.

By 1880, with a maturation of the soil sciences, humans were able to test and evaluate the nutrients in soil and seek more effective ways of increasing its fertility. Soil scientists, in both universities and fertilizer companies, slowly established analytic standards for determining the nutrients in synthetic fertilizers. By then, farmers were buying larger amounts of such fertilizers, which contained mostly phosphates and potash with very limited amounts of nitrogen, the most expensive of the three main nutrients. The earth contains only small deposits of nitrates, located largely in nitrate-rich minerals in Chile or in hardened guano deposits in Peru. The process of creating coke also yields small amounts of nitrate. Until World War I, no one had developed an inexpensive means of extracting nitrates from atmospheric nitrogen. The most common technique involved a cyanamide process that required huge amounts of electricity to gain the free hydrogen needed to create nitrates for both munitions and fertilizer.

Although all soil nutrients have a chemical profile, and the nutrients absorbed by plants are always inorganic in form, the word chemical usually denotes fertilizers with ingredients that humans have somehow transformed from their natural state. Thus, I refer to them as synthetic fertilizers. Historians usually trace the origins of synthetic fertilizers to 1840, when an English chemist discovered that the phosphorus in animal bones, when treated with a solution of sulfuric acid, yielded the types of phosphate that plants could quickly utilize (he called it superphosphate). He gained a patent and began selling phosphate fertilizer in 1842. By 1850, such phosphates, made from bones, were on sale in the United States, but their use was very limited until 1867, when large deposits of phosphate rock were discovered in South Carolina.

As it turned out, the United States had (and still has) large deposits of phosphate ore. Beginning in 1867, phosphate factories in Charleston began using sulfuric acid to convert the rock phosphate into superphosphate. Later, the mining of phosphate ore shifted to other states, notably Florida and Tennessee. The largest market for such synthetic fertilizer was in the Southeast, where soils were severely depleted of nutrients. By 1880, some companies were marketing fertilizer with not only phosphates but also potash imported from Germany and small amounts of nitrates, most imported from Chile and Peru.

Another important ingredient in commercial fertilizers is potassium, in the form of potash (largely potassium chloride). Many soils, particularly clay soils, are naturally rich in potash. Thus, heavily cropped soils often run out of phosphorus and nitrogen before potash. Appropriately, potash is the least critical of the top three nutrients and is usually listed last on standard content labels (nitrogen–phosphorus–potash). Potash was the first valuable industrial chemical exploited in colonial America, and the first patent issued in the United States was for a new method of refining crude potash. But such potash was too expensive to use as a fertilizer until after the Civil War and the development of potash mines in Europe (it is plentiful in dry lake beds or in deposits from long-since evaporated seas). Originally, potash was a direct product of wood ashes, which America had in plentitude. Farmers soaked wood ashes in water to dissolve the potassium carbonate, creating lye. By the eighteenth century, lye, along with fat, was the basis of soap making and was also used in making glass, hominy, and gunpowder. Most farms had an ash pit, which allowed trickling rainwater to create the lye needed for household use. When lye water evaporates, the residue is a form of crude potash, which can be refined into a purer, white powder by heating in a kiln. Although farmers could not afford this early product for their fields, they knew the value of ashes, which they often applied to crops. Also, they recognized the increased yield gained from the ashes of trees burned to create new soil. More than 90 percent of potash from mines is now made into fertilizer, with some mines in the United States but even more in Canada.[9]

Nitrogen depletion long remained the most intractable fertility problem for farmers. Nitrogen is vital for plant growth, particularly for grasses, yet it does not remain in the soil for long periods because of rapid leaching by rainwater. The intense heat of lightning in thunderstorms adds nitrates to rainwater, but not nearly enough for depleted soils. Manure is a good source, but no farmer could collect enough for all his fields. Nitrogen in organic form is available from bones and from plant material, such as cotton seed. Organic decay in fallow fields creates slow-release nitrates. It was largely the nitrogen problem that made it impossible to grow crops year after year on the same land, except for rare soils with deep organic content. Fortunately, in the United States, the rich alluvial soils of the Midwest had enough organic content and enough nitrogen to produce crops for up to a century without fertilizer, but even those soils were eventually depleted. When I was a boy, local farmers desperately needed more manure than their livestock provided, or more clover than they could plow under in the spring. Too often they passed up more expensive fertilizers that included nitrogen and purchased an inexpensive 0–10–4 fertilizer with no nitrogen, resulting in poor yields of corn. They too often piled all their manure on their small but vital tobacco patches. For the garden, or occasionally for tobacco, they bought five-pound bags of Chilean nitrates, or what they called nitrate of soda. After the Depression, the increased availability of affordable nitrogen fertilizer helped launch the agricultural revolution. We enjoyed a nitrogen bonanza.

In 1913 Fritz Haber in Germany discovered a new process for converting the nitrogen in the atmosphere, which is inert and not available for plants, into ammonia (NH_3). Plants can directly absorb the nitrogen in ammonia, making it a valuable fertilizer. The process required a major source of free hydrogen, which was almost always methane (NH_4) the main ingredient in natural gas. The first use of this process was for munitions. During World War I, the federal government constructed the large Wilson Dam at Muscle Shoals on the Tennessee River in northern

Alabama. Its purpose was to gain the power needed to create nitrates, primarily for explosives needed in the war, but with a secondary purpose of producing affordable nitrate fertilizers for the impoverished South. The fertilizer proposals bogged down in the 1920s with the battle over public versus private operation of the Muscle Shoals project, but by then, a few commercial fertilizer companies were using the Haber process.[10]

In 1933, with the creation of the Tennessee Valley Authority (TVA), the advocates of public ownership won out. Muscle Shoals became the site of fertilizer research, development, and demonstration in the National Fertilizer Development Center. In the Depression years, the TVA concentrated on developing new and better phosphate fertilizers, or what it called triple superphosphates. During World War II it produced ammonium nitrate for munitions and started selling it to farmers in 1943. At the end of the war, ammonium nitrate (still widely used) was the first major nitrate fertilizer widely available to farmers. The TVA became the main research and development center for fertilizer, in most cases offering its findings to private companies. It helped develop granulated forms of fertilizer for easier application, developed a very concentrated liquid form of nitrogen (anhydrous ammonia), and created a blend of phosphate and nitrogen (ammonium polyphosphate).[11] For years, the United States led the world in nitrate production, but no more. The soaring price of natural gas has led to the importation of most nitrate fertilizer, as well as most potash. But the United States still exports phosphates.

By 1950 farmers were rapidly increasing their use of nitrates. In many cases, farmers used too many nitrates or used them at the wrong time, with major pollution problems as a consequence. Because of rapid leaching, nitrates remain in the ground for short periods. The most efficient use requires an application at the time of rapid plant growth. Thus, many farmers applied ammonium nitrate or anhydrous ammonia to growing crops, which they referred to as a side dressing. Because of precise measurements of soil nutrients, present-day farmers have often been able to reduce the amount of fertilizer used, with major benefits to the environment. Despite such calibrated use, they now face rapidly rising fertilizer costs, most tied to the surge in the prices of oil and natural gas.

After World War II, all successful farmers used both lime and fertilizer in the quantities needed to maximize yields. Soil testing allowed a calibrated use, with some farmers pushing the use even beyond diminishing returns. The great surge of use came before 1980, not since. From the war until 1980, fertilizer use expanded by around 4.5 percent each year, and this on a slightly declining amount of soil under cultivation. Most dramatic was the increase in nitrogen, from 2.7 million tons in 1960 to 11.4 million in 1980.[12]

This meant that, for the first time in human history, the average farmer could grow crops on the same fields year after year. Previously, this had been possible only on fields that flooded annually, such as in the Nile Valley. The soil has become not the primary source of nutrients but a storage bin to receive nutrients from the outside. In this sense, modern farmers have finally solved many of the problems of soil exhaustion and made most crop rotation a thing of the past, although most corn farmers still rotate corn and soybeans. But note that such continuous cropping can deplete the organic content of soil, increase plant diseases, and thus lower the productivity of soils. Recycling the plant litter each year is only a partial answer. This form of chemical agriculture has other problems, including environmental ones. Yet synthetic fertilizers are essential for the level of production needed to feed the earth's presently inflated population.

Insecticides and Fungicides

In some locations and for some crops (such as cotton during the boll weevil invasion, or corn after the importation of the corn borer), insect infestations are a greater threat than soil exhaustion. More than 700 insects can do serious damage to American crops, fruits, and forests. The use of insecticides dates back to Roman times or earlier. Humans learned how to repel insects with mud baths or the fumes of burning sulfur. By the nineteenth century, several readily available insecticides—sulfur, tobacco juice, soapy water, vinegar, lye, and turpentine—worked for some insects. For enclosed areas, cyanide gas killed most insects. By the twentieth century, three natural insecticides—one inorganic (arsenic) and two derived from plants (rotenone and pyrethrum)—were widely used. I can remember spraying both arsenic and rotenone on our garden crops during the Depression. Both had risks. Arsenic was deadly to animals, including humans, and rotenone to fish. In fact, a major use of rotenone today is to clear ponds of unwanted fish. Some inorganic insecticides remain in use, such as sulfur for mites and chiggers and boric acid for cockroaches. Some of the older, plant-derived insecticides, such as the pyrethroids, are now widely used on field crops.

A revolution in insecticide research began during World War II, leading to hundreds of synthetic (that is, human created) insecticides, although many are derived from existing organic substances. The different families of insecticides are grouped according to the base chemicals used, such as chlorine, carbon, phosphate, nicotine, sulfur, and carbonic acid. The most famous insecticides were the early chlorinated hydrocarbons, led by DDT. In all but one respect (it proved deadly for some species of birds), it was the best insecticide ever developed—inexpensive, broad spectrum, and with no apparent threat to humans. Through mosquito control, it saved millions of lives. Although banned in most countries, DDT is still widely used for malaria control in Africa. Another insecticide in the same family, chlordane, offered almost perfect control over ants and termites but was eventually banned because of its threat to humans.

Another major class of insecticides has a phosphorus base. This class of chemicals includes sarin, a deadly nerve gas. Today, almost all organophosphates are banned. The one exception is Malathion, a broad-spectrum insecticide still used for garden and fruit crops. Some derivatives of nicotine are used as systemics (absorbed by the roots of plants) and for the treatment of turf, rice, and potatoes. One of the most widely used insecticides today—Sevin—is derived from carbonic acid. Organic sulfur compounds are used as miticides. The list goes on and on, but in almost every class of insecticide, a majority are now banned as unsafe for humans or for desirable insects, such as honeybees. It is remotely possible that insecticides or other farm chemicals caused the colony collapse disorder that in 2008 threatened the main pollinator of most fruit and vegetable crops. In other cases, insects developed resistance to certain insecticides. Thus new insecticides appear almost every year, including a recent introduction that controls caterpillars by minute quantities. Despite the risks, insecticides, whether synthetic or natural, are indispensable in modern agriculture, both to maintain the volume of production and to meet the cosmetic standards of consumers (no one wants to find a worm in an apple or an ear of corn).

These synthetic insecticides are not the only means of controlling insects, but in many cases they are the cheapest and most effective. Some bacterial insecticides are widely used, such as Bt (Bacillus thuringiensis), a naturally occurring soil pathogen. It is sold as a powder or a spray and is very effective against the larval stages of moths, mosquitoes, and black flies. It is now incorporated in genetically modified corn to provide built-in insect resistance. One variant is even effective against the boll weevil. It is not toxic to humans, birds, or animals, but it is expensive. Another bacteria that also occurs in some soils is Bacillus popillae, which causes a milky spore disease in Japanese beetles. However, the cost of application is very high, and gaining the necessary level of concentration is always difficult.

Other options include pheromones, which are effective in luring some insects into traps. Some oil sprays and insecticidal soaps protect against certain insects or diseases on fruit trees and vegetables. Alternatively, one can use "friendly" insects to control pests, as when I buy boxes of ladybugs to control aphids on my tomatoes.

Organic farmers make use of all these nonsynthetic forms of control, but at present, it is not possible to control most insects on large fields at an acceptable cost without synthetic insecticides, which are short-lived and in most cases pose no overt threat to human health if one follows all the proper precautions. This is a big "if," because too many home owners and even professional farmers ignore the rules. Everyone who buys fresh food today will inevitably consume some residue from insecticides, even as they benefit enormously from the abundant food supply insecticides make possible. Also, the long-term effects of insecticides on humans may not be known for decades or longer, as indicated by a recent finding that certain inherited problems caused by DDT appear only in the third generation.

Fungal diseases are not as ubiquitous as insects, but they can pose the greatest threat to crops and to human welfare. Human history has been vitally affected by fungi, such as the potato blight in nineteenth-century Ireland. Most crops and many trees are subject to attack from mold, rust, scale, blight, mildew, yeast infection, scab, and smut. Most gardeners are familiar with the fungus problem: roses suffer from black spot, tomatoes from blight, and apple trees from cedar rust. At present, the most active threat to American crops is from Asian soybean rust. In the past, both wheat and corn have faced deadly fungal attacks that placed the whole crop in jeopardy. This has led to a twofold counterattack—the development of resistant plant varieties, and the development and use of fungicides, which can be very expensive when whole crops are threatened. Because diseases can quickly become resistant to a given fungicide, chemical companies develop new ones every year. Today, we have more than 300 listed fungicides in the United States. Most have trade names, and some are very dangerous to humans if not handled with great care. Fungicides have become very specialized, with some aimed at a single disease or a single crop. In many cases, because of the high cost, farmers tolerate a certain level of infection before they resort to fungicides. Also, the pace of plant research is such that resistant varieties are available for most diseases. Note that fungi are not responsible for all plant diseases, such as those tied to nematodes, bacteria, or viruses. In the case of viral diseases, no treatment is available except rotation. Thus my cantaloupes die of fusarium wilt every summer.

Herbicides

Unlike insecticides and fungicides, no organic herbicides were on the market before World War II. Yet few discoveries have so changed agriculture. It all began with chemicals to kill broad-leaved plants (this includes many crops, such as cotton, soybeans, and tobacco). For about a decade, only two existed: 2,4-D and the closely related 2,4,5-T. The latter was Agent Orange of Vietnam infamy, an often deadly gas not because of the herbicide but because of a contaminant in its manufacture—dioxin (since eliminated). Otherwise, these herbicides were safe for humans in all but very high concentrations and remain in use today. Continued research led to whole families of herbicides with different, often highly specialized purposes. Some are useful in water; others kill various grasses. Some work as preemergents, such as those used in yard fertilizer to prevent crabgrass.[13]

In crop agriculture, herbicides were revolutionary for one reason: they all but eliminated the need to cultivate row crops. This meant an enormous saving of human labor and a dramatic increase in production per acre. For corn, herbicides raised production more than had hybridization. Farmers could now reduce the width of corn rows from three feet or more to as little as twenty inches, in some cases almost doubling production. With the introduction of preemergents in the 1980s, cotton farmers could do the same. Thus, by 1982, American farmers used herbicides on 95 percent of corn and 93 percent of cotton.

In the 1980s new nonselective, or universal, herbicides (they kill all plant life) made possible what is called no-till cultivation. Instead of plowing and harrowing the fields, one simply applied a universal herbicide at a safe period before planting, leaving a cover of dead vegetation. No-till has been almost universally adopted in some regions of the country, particularly for row crops. In certain river basins it has become a near necessity to meet erosion or pollution standards, often with Department of Agriculture subsidies available to help make the transition. In one final strategy, Monsanto scientists are now genetically modifying soybeans to make the plants immune to the most popular nonselective herbicide—its own Roundup.

No-till required new and, in some cases, expensive planters of varying design. My brother-in-law used a no-till corn planter with a deep, single-foot plow, like one used for subsoiling, to break through the hardpan created by years of normal plowing. Following this were two disks that created the furrow for seeds and two funnels for fertilizer, followed by a covering wheel. In the hilly Palouse of Washington, wheat farmers developed a heavy drill with enough weight to allow the row-creating disks to penetrate the stubble. These costly machines require a huge, eight-wheeled diesel tractor to pull them. Corn farmers can now plant in one operation and, other than the follow-up application of a selective herbicide, do nothing more until they combine the corn in the fall. No-till does not necessarily increase production, but it saves labor, protects against erosion, and, in critical watersheds, lessens the chemical runoff. Because it delays the drying of the soil in the spring, no-till is not practical in some northern states, where farmers often substitute low-till methods (with disks or chiseling plows) or ridge-type cultivation.[14]

Herbicides pose some dangers. They can cause problems when broad-cast from the air, for even minute particles can damage nearby crops. When used to control vegetation along roads and railroads, they dramatically reduce wildlife habitat, as Rachel Carson emphasized. Thus, great care is needed in their application. In order to purchase some herbicides, farmers must obtain a license, which is issued only after they attend courses on methods of safe application.

Antibiotics and Steroids

A final category of chemicals relates almost entirely to livestock. Antibiotics and hormones have become both important and controversial. Just after World War II, penicillin, considered a miracle drug in military medicine, became the first great antibiotic. Even as humans benefited from its use, veterinarians began prescribing it for farm animals, with great success. Such therapeutic uses of penicillin and other antibiotics became routine. Many farmers learned how to store and use these drugs without having to call a vet. Antibiotics had a significant impact on livestock health and productivity. Chronic diseases, such as mastitis and pink eye in cattle, were finally curable. Without antibiotics, the hog and chicken operations of today would have been impossible. Chicken coops housing up to 20,000 birds, and narrow boxes or cages holding up to 4,000 hogs, create high stress levels and invite epidemics.[15]

It was the nontherapeutic use of both antibiotics and hormones that created decades of continuing controversy. Both clearly enhanced growth and thus led to dramatic gains in meat and milk production. Farmers began routinely mixing antibiotics with animal feed not to treat any illness but to prevent infections and, even more important, to stimulate growth or allow a more efficient use of feed. The growth-enhancing effect of antibiotics for healthy animals was a surprise, and one that has not been fully explained. Somehow, a steady diet of low-level antibiotics kills harmful bacteria in the stomachs of cattle, hogs, and chickens and thus makes the digestive process more effective. By the 1980s, the use of such antibiotics had become standard practice in poultry and hogs and, in some cases, feedlot beef cattle.

By some estimates, up to 70 percent of all antibiotics in America are consumed by livestock. The U.S. Food and Drug Administration (FDA) has approved all the antibiotics in use (up to twenty-five in all). The drugs, which are administered in very low concentrations, must be withdrawn from animal feed well before slaughter, and the FDA has set strict standards for antibiotic residues and has mandated tests to ensure compliance. Even so, this practice is now under attack in all meat-producing countries. Concerns largely involve the possible development of antibiotic-resistant strains of disease-causing organisms, as well as issues of animal welfare. Most of the debates center on poultry and the possibility that humans might be infected with resistant strains of *Salmonella* from chickens.

By 2002, the tide had clearly turned against drugged chickens. The major chicken producers, including Tyson and Perdue, announced the end of the use of antibiotics for growth enhancement, with a bit of hedging about their use for disease prevention. In 2007 a carefully crafted bill in Congress sought to ban most growth-enhancing antibiotics, a step already taken in European countries. McDonald's led the way among fast-food companies in rejecting drug-treated chickens. Meanwhile, the widespread agitation helped create a rapidly growing market for drug-free chickens. Animal welfare issues helped boost the sale of free-range chickens. Thus, even though antibiotics helped make large chicken factories possible and increased the birds' weight gain by up to 10 percent, they never really revolutionized this industry, although their role in controlling diseases remains critical. Since hog factories use the same antibiotics, they face the same pressures to drop them. For cattle producers, who never used antibiotics routinely, the controversy largely involves growth-enhancing hormones.[16]

In the case of beef cattle, hormone use has not aroused the concern of most consumers (an early use of hormones in chickens was subsequently banned). But critics have raised enough concerns to create a growing market for organic beef or at least beef without any hormones. In large cattle feeding lots, sex hormones or, more often, synthetic derivatives (beginning in 1951 with stilbestrol, a synthetic estrogen) are routinely implanted in the animals' ears in the form of small pellets. The preferred drug today is Trenbolone. The result is comparable to steroid use by humans. The hormones bulk up the muscles of cattle and thus increase the amount of lean meat by 15 percent or more. Since European countries have banned such hormones, one effect of their use has been a loss of some export markets. The hormones increase, very slightly, the level of sex hormones in the resulting meat, but this apparently poses no risk to humans. Some critics disagree with this benign judgment, and a growing segment of consumers has expressed concern. To outlaw such hormones would be a major blow to large beef producers and would also deny American consumers the type of lean beef they now prefer.

Dairy farmers have their own special chemical—recombinant bovine growth hormone (rBGH)—but they may have to give it up. Here the controversy has been intense and often marked by more heat than light. Monsanto, which developed and patented rBGH, is also responsible for the development of genetically modified field crops. Because this hormone occurs naturally in all cows, Monsanto claims that it is not dangerous. In the past, small amounts of the hormone had been taken from the bodies of slaughtered cows and used to stimulate milk production in special situations. But this source was so limited as to preclude general use. Monsanto scientists found a way to use recombinant DNA to produce large quantities of rBGH, using the same techniques that had led to a synthetic form of insulin. In 1993, in one of the most controversial decisions it ever rendered, the FDA approved the use of rBGH (Monsanto called it Postlac) for milk cows. Monsanto correctly argued that the hormone was chemically similar to what cows produce themselves. The added quantities stimulate lactation and thus increase the milk produced by a given cow by up to 15 percent. For large dairy operations, this could mean the difference between profits and losses. Thus, 30 to 50 percent of all milk cows (the estimates vary widely) were on Postlac by 2007. Because of the cost, it is not widely used by small dairy farmers.

The controversy over Postlac first climaxed after its introduction in 1993. The one critical issue was whether the milk was safe for humans. Like any drug, rBGH posed some risks for the cows. It increased the incidence of mastitis, the leading udder disease in milk cows, partly as a consequence of the increase in lactation. For this reason alone, European countries and Canada refused to approve its use. One consequence of an increased level of BGH in the blood is an increase in another growth hormone, insulin growth factor (IGF-1). Whereas BGH is distinctly different from human growth hormone, IGF is almost identical in both species. Thus, milk containing BGH could increase human IGF, although the amount would be very small. Is this dangerous? Many activists insisted that it might be, and they gained support from concerned scientists. The controversy seemed to lessen in the years after 1994, only to revive again after 2002, and as I write, it is still ongoing. Images of a monopolistic company, inflated costs for patented medicines, huge dairy farmers driving smaller ones out of business, and the array of new genetically modified food products added to the emotional level of the debate.

Monsanto fought to protect its image and its profits. It sued a small New England cooperative that advertised no BGH in its milk. Monsanto argued that this claim constituted false advertising, since all milk had BGH, not just that from cows given Postlac. It won, and the cooperative had to note on subsequent ads that there was no evidence of any difference between milk with natural BGH and that with rBGH. After this decision, companies that wanted to emphasize the absence of this synthetic hormone had to use careful and somewhat evasive language. Yet the opposition to rBGH grew, and organic dairies flourished. The tide seemed to turn decisively in 2007 when both the Safeway grocery chain and Starbucks coffee shops announced at least partial bans on rBGH milk. It seems likely that other companies will join the crusade, and dairy farmers may have to abandon the use of rBGH to preserve their market. Some cooperatives and some local dairies have already made the shift. Whether the FDA will withdraw its approval is more doubtful.

PLANT AND ANIMAL BREEDING

A final necessary condition for the agricultural revolution also involves genetics, in the sense of selective breeding of new varieties of crops and livestock. The largest breakthrough was hybrid corn, which was developed in the 1920s and gradually accepted by farmers in the 1930s. It is arguably the most important plant-breeding innovation in the United States.

The development of hybrid corn goes back to 1907. Two scientists began experiments that involved the crossing of two strains of inbred corn. They proved that such crosses led to greater vigor. The emerging ears of corn are pollinated by pollen from the tassels of its own or nearby stalks (or, on a windy day, quite distant stalks). This results in a diverse genetic mix in any open-pollinated corn. Some have referred to such corn as a medley of past unplanned hybridizations. By isolating a single stalk of corn and allowing only self-pollination, one can reduce the genetic diversity. By continuing to self-pollinate over several generations, one can isolate certain desired traits that were hidden in the larger gene pool. However, such inbreeding, and the loss of alleles and thus genetic diversity, almost always leads to less vigorous plants. The first hybrid corn resulted from the crossing of two selected inbred strains, which in some cases led to a desirable outcome—the preservation of desired traits combined with renewed vigor. Donald Jones, the one person largely responsible for the development of hybrid corn, took this a step further. At his Connecticut experiment station, Jones crossed two of these hybrids, creating a double-cross hybrid. The inbred lines that produce a desired hybrid are a valuable commodity and are eligible for patent protection. Since hybrid varieties do not breed true, those who own such patents have a valuable commercial product. Seed companies thus gained a monopoly on hybrid seed corn. For years, hybrid seed production required a good bit of hand labor, as workers had to cut off the tops of the female plants. Today, a form of cytoplasmic DNA, bred into female plants, renders them sterile, eliminating the need for topping.

As early as 1922, Secretary of Agriculture Henry C. Wallace supported hybrid research in his Bureau of Plant Investigations. Soon, several experiment stations were involved in hybrid research. Back in Iowa, Henry A. Wallace, the son of Henry C. and editor of an influential agricultural periodical, carried out hybrid research on his own and in 1925 formed a new company to sell hybrid seed—the Hi-Bred Corn Company. In 1925 the first private farmer planted his fields with hybrid corn. On fertile soil, the early hybrids increased production from 10 to 15 percent. At first, farmers were slow to adopt the hybrids. Some tried to save seed and lost all benefits. Others, with poor soil, did not gain larger yields. But gradually, during the Great Depression, a majority of Corn Belt farmers began buying the more expensive seed (78 percent used it by 1949). The one deterrent to boosted yields turned out to be low levels of soil nitrogen, which was cured by the revolution in nitrate production after the war, as described earlier. One of the desired traits bred into hybrid seed was a stronger stalk, which, by design, facilitated the machine harvesting of corn, also described earlier (only 10 percent of corn was harvested by machines in 1938).[17]

After the success with corn, hybridization spread to most vegetable crops. In addition, plant breeders, working in federal experiment stations or in seed companies, continued to select desirable traits generation after generation, leading to the introduction of new varieties every year. In some cases, the goal was disease resistance, as well as a heavier yield or better shipping qualities; all too rarely was improved flavor the objective. Such improvements also occurred in all orchard crops.

These rapid genetic improvements in major crops had an impressive international influence. In 1941 the Rockefeller Foundation established a new research center near Mexico City (it still exists) to seek improvements in corn and wheat varieties, particularly those suited to underdeveloped countries. This research soon expanded to other crops, such as rice. Norman Borlaug, who developed high-yielding wheat varieties at this center, received the Nobel Prize for his achievements. The spread of such new varieties, along with the upsurge in the use of fertilizers and chemicals, had a major impact not only in Mexico but also in other subtropical or tropical countries, particularly in northern India. Soon, almost everyone was referring to a "green revolution." Contrary to Borlaug's hopes, the greatest impact occurred among large, highly capitalized farmers, not among the small peasant farmers who could not afford the capital inputs necessary for the best use of the new crop varieties.

The gains in livestock breeding were as impressive as those for crops. The greatest improvement was in milk output, which more than doubled from 1950 to the present. Shifting consumer demands (such as for leaner hogs and lower-fat milk) led to shifts in favored breeds and, in many cases, a trend toward one or two dominant breeds in each livestock group. For example, Holsteins, larger and heavier milkers but with less butterfat, have won out over Jerseys and Guernseys among dairy cows, while Angus seems headed toward domination among beef breeds. The single greatest innovation in animal breeding has been artificial insemination, which allows a relatively few selected males to generate thousands of offspring. Perhaps next in importance has been the development of detailed production records. In the future we will gain detailed DNA profiles of breeding stock. Before World War II the emphasis was on pure breeds, but once the breeds were well established, the goal for most farmers was not breed purity but the highest level of production, which sometimes favored mixed breeds. In beef cattle, this has led to the importation of European breeds and Brahman cattle from India.

This completes a summary of four major developments that helped ensure the rapid growth in American agriculture after World War II. Although the growth was most revolutionary between 1950 and 1970, the level of productivity has not slowed to any appreciable extent since 1970. But the pace of change has slowed in several areas. The number of farms has remained relatively stable, and so has the number of farm operators. The amount of land in cultivation is stable or declining slightly. In farm machinery, no innovations have matched the self-propelled combines and cotton pickers that matured in the 1950s and 1960s. Structurally, despite widespread opinions to the contrary, American agriculture is still dominated by family-owned farms, although in some cases families form partnerships or corporations in which the family retains majority ownership. In fact, from 1997 to 2002, the number of investor-owned farms declined. Nothing suggests any change in ownership patterns in the near future.

Yet, in the early twenty-first century, several trends reveal an increasingly closer tie between farmers and the agricultural businesses that process, transport, and sell their products. At times, the boundary between the farmer and the processor has blurred. More and more vertical integrations have developed in several crop and livestock areas, with poultry leading the way. This has steadily diminished the managerial role of farm owners and operators, even as it has lessened their risks. When farmers contract with a company to raise a certain number of chicks (supplied by the company) according to certain specifications and to sell them when mature at a negotiated price, the farmer is as close to being a company employee as an independent operator. In many vegetable and fruit crops, and increasingly in tobacco, contract farming is well established and is growing in almost all areas. This means that the family farm survives, but with a drastic shift in what that means.

One other new development—genetically modified crops and even livestock—may or may not be as revolutionary as earlier innovations. If such crops continue to gain acceptance from farmers and from skeptical consumers, farmers will become even more dependent on seed suppliers. At present, more than 40 percent of American corn includes Bt to control insects, and herbicide-resistant soybeans are becoming the norm. Thus, genetically modified crops may launch a second green revolution, in the sense that they may increase output, eliminate the need for many pesticides, and increase food's nutritional value. These changes are very controversial, particularly outside the United States. As yet unidentified dangers may limit their expansion. But it is at least possible that new leaps in productivity lie ahead.

6. Surpluses and Payments
Federal Agricultural Policy, 1954–2008

In many respects, the agricultural revolution after World War II was a blessing. Consumers were able to spend a steadily decreasing share of income on food and fiber. The diversion of workers from agriculture freed up labor for manufacturing and services, leading to a higher growth rate in the overall economy and to a level of consumption undreamed of in the human past. One of the best indicators of a nation's overall prosperity is the percentage of workers required to feed a population—more workers in agriculture usually mean a lower standard of living, unless a large share of that product is profitably exported. But as in any revolution, there were losers, including many who were forced to leave farming and make a difficult adjustment to some other occupation. Even the winners, those who continued to farm more efficiently with each passing year, did not always gain a fair return for what they contributed to the nation as a whole. This leads to the thorny issue of farm prices and farm incomes.

PRODUCTION CONTROLS AND PRICE SUPPORTS

The statutory authority for federal involvement in the marketing of farm crops remained in place through the Korean War. With a few exceptions, such as tobacco, the acreage controls, price supports, and marketing agreements were unneeded until at least 1948, and for most crops not until the end of the Korean War. A devastated Europe and Japan ensured a strong demand for American farm goods until a gradual recovery began in the early 1950s. By then, an unprecedented surge in production was taking place on American farms. From 1954 to 1972 farmers consistently produced more crops than they could sell. Surplus became a ubiquitous term and concern in these years, as Congress struggled to find an answer to the "farm problem."

In a productivity revolution, it is always difficult to generate enough demand to absorb an avalanche of goods. This is most true of food, where demand is inelastic. Until the end of the Korean War, exports and the baby boom kept demand in line with supplies, with the exception of feed grains in some years. The rate of population growth continued through the 1960s, but it could not match the growth rate in agriculture. A more important factor in causing price instability was shifts in foreign demand.

One may ask, as economists do, Why not let the "market" control prices and the allocation of labor and capital to a given sector or industry? Of course, in a strict sense, the term *market* is a bit of a mystification, for there is no market actor making decisions, but only various individuals or groups exchanging labor and goods. They can safely do this only in a polity that protects property ownership, ensures the fulfillment of contracts, enforces rules to uphold the common welfare, and uses taxes to provide necessary public goods such as law enforcement, defense against foreign enemies, education of children, and protection of the environment. Within this context, one may argue that the best agricultural policy is to let individuals decide what to produce, just as other individuals decide what farm products they want and are willing to pay for. If too many farmers grow corn, prices will certainly fall, given the inelasticity of the demand for corn. Low prices will encourage farmers to grow less corn, shift to other crops, or even give up on agriculture altogether and take up another occupation. Prices will rise as corn production slows, and over time, the interaction of supply and demand will lead to an equilibrium in which corn prices and the number of corn farmers remain close to a "rational" or "market-determined" level, with optimal benefits to society as a whole.

Until the New Deal, federal agricultural policy generally adhered to such a market model. In fact, agriculture in many respects remained closer to this ideal than did other sectors of the economy. It had no monopolies or restraints on trade. More than in any other sector, agriculture was made up of millions of individual producers competing with one another. As the most basic economic sector, the one most vitally involved with human welfare, agriculture had received plenty of support from state and federal governments, but without any direct control over agricultural marketing. It had seemed to be in the national interest to open up public land at low cost, provide free homesteads, take over the costs of agricultural research and development, provide credit opportunities, and facilitate cooperative marketing, yet without exerting any direct control over farm management or farm marketing. To justify this infrastructural aid, one could cite the long history of tariff protection for manufacturing, large land grants to railroads, and limited liability rights for corporations. Unlike the corporate enterprise that was dominant in manufacturing, millions of farmers could not coordinate policies or exert much control over marketing and prices, except by using their tremendous political clout to gain desired policies from Washington.

The New Deal commodity programs reflected a response to depression. It was not clear that such emergency legislation would fit the postwar economy. In fact, farmers as well as legislators hoped that something close to a free market would be possible. Farmers wanted the freedom to grow what they wanted. Consumers wanted low prices for food. The Agricultural Adjustment Act of 1938 remained the basis for all types of production controls and price supports, but Congress amended it almost every year. Its primary policy, whatever the implementing details, endured at least until 1996. To control production of the major storable commodities, it allowed the federal government, working through the committee system in each county, to establish either voluntary acreage allotments for specified crops or mandatory acreage or marketing quotas, based on past levels of production. The voluntary acreage allotments meant that a farmer could sign up or contract to set aside a specific number of acres for a controlled crop. If he adhered to his contract, as certified by a local committee, he was eligible for a nonrecourse loan at a set price level, usually expressed as a percentage of parity. Those farmers who did not sign up for acreage limits were usually ineligible for such loans and thus had to sell at market prices. However, they were free to plant and sell all they wanted, an option that often appealed to very large and efficient farmers or those who opposed the government commodity programs and campaigned for a free market in agriculture. In a sense, the price support was a bribe to get farmers to produce less during periods of surplus and declining prices. Note that in almost all cases, the price supports meant prices above the international level, which required routine tariff protection for American farmers.

The mandatory quota system applied only to crops in which a two-thirds majority of farmers voted in favor of it. This was the only system ever used for the two major types of tobacco, for peanuts and rice in most years, and for wheat from 1954 to 1961. The market for soybeans soared by the 1960s, but this exportable crop was in such high demand that only stabilizing crop loans (at a rate well below competitive price levels) were offered before the farm bill of 1985. Most perishable vegetables and fruits, as well as livestock, did not fit a price-support system based on nonresource loans. Notably, the farmers who diverted acres from controlled crops were always prohibited from shifting these acres to fruits and vegetables.

Farm Policy in the Truman and Eisenhower Administrations

Farm policy became a hot issue in the Truman administration. The often bitter debates led to no major policy changes but offered an early rehearsal for later programs. Early in World War II, to encourage farmers to produce as much as possible for the war effort, Congress so modified the Agricultural Adjustment Act of 1938 (by the Steagall Amendment) as to guarantee farmers 90 percent of parity on not only five basic, storable commodities (corn, wheat, rice, cotton, and tobacco) but also on thirteen other perishable products, including milk, hogs, soybeans, peanuts, chickens, eggs, potatoes, and sweet potatoes. This price support was to remain in effect until two years after the official end of hostilities, which turned out to be the end of 1948. In only one case did this commitment increase farm income, because almost all agricultural prices remained above parity levels until 1949. The exception was potatoes. Higher levels of production and decreasing demand (the age of fast food and french fries had not yet arrived) led to huge surpluses just after the end of the war. Since potatoes cannot be stored for an extended period, the Commodity Credit Corporation (CCC) had to buy the surplus and pay farmers 90 percent of parity. In three years it lost $80 million on rotting potatoes, some of which it burned. With postwar inflation and rising food costs, consumers considered this a major scandal. It was an early warning of what would happen to many crops in the 1950s, but after 1948 crop loans were at least rarely available for perishable crops.

In 1947 Congress began extended debates on what to do about the support programs after the expiration of the Steagall Amendment. Without a full consensus, it first voted to extend the 90 percent guarantee until the end of 1949. But with broad support from all major farm organizations, and seemingly from the Truman administration, it voted to begin a program of flexible price supports for the major storable crops in 1950, with the range of support from 60 to 90 percent of parity. When surpluses developed, the support level would drop. It was assumed that this would lead to a decrease in production. To gain the nonrecourse loans, farmers would have to accept production limits. Only the flexibility feature was new. For perishable crops, Congress approved, at least in principle, production payments to farmers who grew certain crops such as potatoes. These payments would amount to direct subsidies, since taxpayers rather than consumers would have to provide the funds. This was not completely new, for grade B milk producers—primarily for butter, cheese, and dried milk—had sold surpluses directly to the government during the 1930s.

Many farmers accepted the idea of flexible supports, but not most cotton farmers. By 1949, as prices weakened and crops such as corn faced surpluses, most farmers clearly wanted the security of fixed supports. Even though President Truman had signed the bill providing for flexible supports, it was clear by the 1948 election that it would be politically expedient to amend the existing law. Secretary of Agriculture Charles Brannan finally worked out what seemed, to many, a radical new policy for commodity programs. Ultimately, after bitter debate, Congress in 1949 and again in 1950 rejected the Brannan Plan and simply extended the 90 percent parity through 1950, with new, flexible supports to be gradually phased in over three years. Before the new plan could go into effect, the Korean War began, and demand for agricultural commodities quickly soared. Congress responded by continuing the existing 90 percent parity through 1954. At that point, the Eisenhower administration, guided by Secretary of Agriculture Ezra Taft Bensen, finally implemented a flexible support scheme, which proved to be very unpopular with farmers.

What made the Brannan Plan so controversial? What he proposed, in retrospect, does not seem as new or revolutionary as critics claimed at the time. He wanted to update the base period for determining parity (previously 1910 to 1914) to the ten years from 1939 to 1948. As a bid for farmer support, he asked for a continued crop loan program at 100 percent of parity for corn, wheat, cotton, and tobacco, conditional on conservation practices and acceptance of acreage or marketing controls whenever needed. He asked for direct production payments for producers of milk, eggs, chickens, hogs, beef cattle, and lambs. These payments would make up the difference between market prices and prices that assured a fair or parity income to producers, or what would later be called target prices. He did not base eligibility for such payments on any required reductions in production but only on support for conservation measures. Note that these payments would be funded by taxpayers, and they would allow prices for these vital foods to reflect the interaction of demand and supply, or what promised to be lower food prices for consumers facing a rapid increase in most other prices. Finally, he wanted a cap on the amount of commodities eligible for payments of any type; for most crops, this was the amount that would have sold for about $25,000 (approximately $200,000 in 2007 dollars), or roughly caps on benefits for 2 percent of all farmers. He hoped such caps would impede the growth of large industrial farms and thus help preserve family farms. He offered no special aid to threatened small farmers but hoped to implement some program for them in the near future.

The Brannan Plan involved a carefully developed program to appeal to both farmers and consumers. What was unclear was the probable cost to taxpayers. Resistance focused on cost, the shift of benefits from consumers to taxpayers, the emphasis on farm income rather than commodity prices, and the limits on payments. In the years of anticommunist agitation, a developing cold war, and intense political partisanship, the details of the proposal were no more important than the ideological and class issues. Brannan, a former official in the Farm Security Administration and a close friend of James G. Patton, president of the National Farmers Union, seemed too far to the left politically. During the fight over the Brannan Plan, the American Farm Bureau, the largest organization of farmers, opposed the plan and came to despise Brannan. In retrospect, it is clear that the plan would have cost more than the public was willing to pay, offered farmers too much in the way of special protection and direct subsidization, and included too few strategies to reduce production. But it is important to note that as early as the Kennedy administration, support for farmers shifted from an artificially high cost for commodities to direct subsidies from taxpayers. Even today, target prices and deficiency payments have echoes of the Brannan Plan.[1]

The clash over farm policy revived at the end of the Korean War. But never again did farm bills attract as much public interest as the Brannan Plan had, when even high school debate teams made this a popular topic. Year after year the controversy raged over the level of support (a percentage of older parity formulas or new target prices), which crops to include, and how to fund the costs of storage and recoup the loss from forced sales.

With wartime supports due to expire in 1954, the Eisenhower administration tried to move back toward a free market for agriculture. In this effort, it had the support of the Farm Bureau. In a major farm bill in 1954, set to take effect in 1955, Congress enacted a flexible price support plan. For 1955, for most storable commodities except tobacco (rigid acreage controls allowed it to retain 90 percent of parity, with no major expense to taxpayers), the support price could range from 82.5 to 90 percent, and in the following years from 75 to 90 percent. Perishable crops did not receive any loans, and only wool growers would receive production payments. These flexible supports quickly became unpopular among farmers, as surpluses mounted and prices remained well below 90 percent of parity. Government warehouses were soon overflowing with stored commodities (at the highest level in history by 1960). Low loan rates (often only 75 percent of parity) failed to slow production. As an added strategy for quickly reducing surpluses, Congress and the Eisenhower administration opted in 1956 for a program that harkened back to the land purchase program of the 1930s—an Acreage Reserve Program (popularly called the Soil Bank) that lasted for only three years, during the crop years 1956 to 1958.[2]

The Soil Bank allowed farmers in six major crops then under acreage control and price supports to sign contracts to take some, or in 1958 up to all, acres in a given crop out of production for a year. This land had to remain idle. They received what the government calculated to be the net income they would have gained from cropping the land—a generous offer. For example, they received, on average, $40 to $44 for each acre of corn land retired, with the actual amount based on location and past production records. By 1957, farmers idled more than 21 million acres, or around 2 percent of all crop- and pastureland in the country. In the final year, 1958, participating farmers could not expand the acres planted in other crops, a policy that caused many to drop out of the program. The Soil Bank undoubtedly lowered production, but productivity improvements in agriculture were so rapid that overall yields continued to rise in all but one year. This, plus farmer opposition, doomed the program. In some cases, the leasing of land to the government facilitated the movement of marginal farmers out of agriculture.

A second program, the Conservation Reserve, combined conservation with land retirement. Farmers could enter into contracts to retire cultivated land, whatever the former crop, for three to ten years. If the contract included tree planting, it could extend to fifteen years. Farmers received a low rent for the land and extra payments for various conservation measures (tree planting, holding dams, wildlife protection). A farmer could lease all his land (this turned out to be the favorite tactic after 1959, when a total lease gained a special bonus), but after 1958 the total annual rent on any farm could not exceed $5,000. This reserve eventually included more than 28 million acres, but most of the retired land had been marginal farmland. Thus, the program did little to reduce crop surpluses. The government offered the last contracts in 1960. Only in 1985 did Congress reenact a modified version of this program, in this case only for highly erodible land.

Ultimately, the Eisenhower administration found no acceptable solution to the problem of surpluses and low prices. In part, this was because many large, efficient farmers did not participate in the voluntary loan program. Even those who did, and thus reduced their acreage by up to 20 percent, compensated by raising more crops on their remaining land. Thus, the main commodity programs raised prices above world market levels, which hurt consumers, yet failed to please farmers. In 1959 the Department of Agriculture dropped all production controls over corn and set the loan rate at the market level. Production soared. So did surpluses. Prices plummeted.

Managing Surpluses during a Productivity Revolution

In 1961 the Kennedy administration began a new program that eventually applied to almost all storable crops. For feed grains (primarily corn, but also a rapidly growing amount of grain sorghum), it revived acreage controls, required that farmers devote diverted acres to conservation practices, and set the loan rate (the price support level) close to the market rate. To compensate farmers for this low level of support, the government offered all contracting farmers acreage diversion payments. In a typical year, if a farmer diverted 20 percent of his normal corn acres (based on the level of production for the last two years), he could receive half the normal income from the unplanted acres. In some years, for a 40 percent reduction, he could receive 60 percent of normal income. Small farmers with fewer than 25 acres could divert their whole crop and receive 50 percent of normal income. This popular program enticed most corn farmers to sign voluntary contracts. Note that this shift had important consequences. It came close to keeping corn supplies in balance with demand. More important, it allowed the market price of corn to remain low, a valuable benefit to consumers and, in particular, to beef and hog producers. Yet it did not lower the income of farmers below previous levels because of the generous diversion payments. The cost of these payments came not from consumers but from taxpayers. Within five years, the amount of crops in storage dropped to one-fifth the level of the mid-1950s.

For the two major export crops, wheat and cotton, the new system was much more complicated. In 1961 both were subject to mandated production controls and eligible for crop loans. This remained true for cotton, but wheat farmers, in a well-publicized referendum, rejected any mandatory controls in 1963. However, most wheat farmers signed up for voluntary crop reductions the next year. These contracts involved low loan rates, but the farmers received two levels of diversion payments, based on the proportion sold in the domestic market (higher) and that exported (lower). Tobacco, as always, remained under its tightly controlled acreage or marketing quotas, with loan rates and prices closely aligned. The CCC continued to support the price of grade B dairy products by direct purchase on behalf of both domestic and foreign food aid programs, but the rate of support was generally well below parity. Note that the new approach (low loan rates and diversion payments to maintain farm income at close to parity) was still based on production records, not need. This entailed large diversion payments to the largest and most efficient farmers. It also provided an incentive for farmers to maximize production on their nondiverted acres, to use more fertilizers and pesticides, and to increase their base acreage by lease or land purchases, thus forcing small farmers out of the market. During the Nixon administration in 1971, in an attempt to deal with equity issues, Congress limited total diversion payments to each farm operator to $55,000, the first of a series of caps that never worked as intended because of various clever strategies to circumvent them.[3]

In 1969 most farm prices, led by wheat, began to rise. Within two years, American farmers moved into a rare period of high exports and surging prices, in the midst of an inflationary crisis. A major grain deal with the Soviet Union in 1972 quickly used up accumulated surpluses, and wheat prices were soon at record levels. By 1974 the CCC had almost no crops in storage. In 1974–1975 this boom reached a climax. People everywhere were concerned about what they believed would be a severe and enduring world food crisis. Most acreage and support programs were no longer needed. In fact, wheat stocks fell so low that the government resorted to incentives to increase production. Happy farmers borrowed money, bought new machinery, expanded landholdings, and looked forward to a decade of prosperity. And even though prices leveled off and then declined after 1975, the collapse was less severe than in 1921. Farmers were better organized, however, and in 1977 they held protest marches in Washington, D.C. Their anger dissipated a bit when prices and exports recovered in 1978 and 1979. Unfortunately, the worst was yet to come.

One new strategy marked the commodity program after 1973. Instead of support prices based on some percentage of parity, the new farm legislation required Department of Agriculture economists to set annual target prices for each crop, based on expected costs of production. The targets were almost always above the loan support level, and when market prices were below targets, farmers received direct payments to make up the difference. When loan rates were above world market prices, farmers enjoyed a double subsidy—gains from nonrecourse loans (paid for by consumers) and deficiency payments (paid by taxpayers). But low loan rates meant that farmers rarely gained much from unredeemed loans. Instead, they relied on deficiency payments to gain the targeted income. These payments went to both farmers who took out loans and those who did not. Such deficiency payments remained a part of agricultural price supports until 1996, and in the form of counter-cyclical payments they still exist today.

THE FARM CRISIS OF THE 1980s

A deep farm depression that closely resembled the one in 1921 marred the first half of the 1980s. The new Reagan administration was committed to moving back to free markets in agriculture. But at times, pressure by Congress and by a powerful farm lobby forced it to do just the opposite. Payments to farmers rose to four times their earlier levels, and surplus commodities held by the CCC rose to their second-highest level. By 1985, the cost of the total farm program soared to more than $20 billion (approximately $40 billion in today's dollars), or what many people saw as a national scandal. But such was the nature of the program that it did not raise the cost of food. In fact, the collapse of farm prices, to a limited degree, lowered food prices, since the payments to farmers came directly from tax dollars.

In 1980 almost every conceivable factor worked against foreign demand for American farm products. In January President Carter placed an embargo on wheat exports to the Soviet Union, one of several responses to the Soviet invasion of Afghanistan, and an action that helped elect Ronald Reagan in the fall. Interest rates soared in the midst of rampant inflation, in part caused by a major oil crisis, while an overpriced dollar deterred purchases of American crops. A developing world depression led to a steady drop in grain exports. In five years, from 1980 to 1985, total farm exports fell by half. In 1980 alone, farm income fell by 46 percent, a faster rate than during the Great Depression. Land prices, which had risen rapidly in the booming 1970s, fell sharply from around $1,200 an acre in 1980 to less than $700 in 1988. Exuberant farmers, many of whom had recently purchased new self-propelled combines or bought out neighboring farms, were suddenly faced with bankruptcies. Sixteen percent of commercial farmers were financially stressed by 1985. The income support programs and federal loans helped most farmers avoid bankruptcy (only about 2,100 filed for bankruptcy in 1985–1986), but up to 300,000 left farming and, in many cases, sold their land. Once again, surpluses grew, and so did the storage and disposal problem for the CCC. Dairy stocks became so large by 1985 that the CCC agreed to purchase whole herds of cows at market price and commit them to slaughter if owners would promise to halt dairy production for five years.[4]

In 1983 Congress approved a new tool to get rid of CCC inventories: it offered a one-year emergency payment-in-kind (PIK) program to farmers. If they contracted to reduce acreage by a given amount, they could receive certificates, redeemable by the Department of Agriculture, that documented their ownership of so many units of a controlled crop, in amounts up to 80 to 95 percent of their normal yield on the diverted acres. Most of the surplus was in local warehouses or elevators leased by the CCC. Farmers greeted the program with unexpected enthusiasm, for it allowed them to gain corn or wheat without all the work of farming. They diverted 82 million acres from production, or more than a third of all cultivated acres in program crops. Aided by a widespread drought, surpluses declined for one year at a rapid rate, and most farm prices temporarily rose. As farmers sold their certificates, even exports rose. But the cost was very high, since the CCC gave away a huge amount of its assets (stored commodities)—an estimated $10 billion to $11 billion worth. Since these costs did not come directly from the Treasury, they did not show up in those attributed to commodity programs, which reached a near-record $18.5 billion. This open-ended PIK program lasted for only one year, but the CCC continued to offer farmers generic PIK certificates (they could choose any commodity with a large surplus) as an alternative to normal monetary deficiency payments, with an added inducement for very large farmers (PIKs did not count toward the cap on total payments to any one operator).[5]

At the climax of this farm crisis in 1985, Congress enacted a new farm bill. It modified existing crop loans and deficiency payments, added some incentives to get farmers to redeem loans early, raised some target prices, increased acreage reductions on most crops (up to 27.5 percent for wheat), and added to a growing number of export subsidies. Because of new provisions, the former nonrecourse loans were now called marketing loans. They were so devised as to eliminate most storage of crops by the CCC. Farmers could still take out loans, but at rates close to world prices. Such loans guaranteed a floor price, and farmers could still forfeit such loans at the end of the nine-month loan period. But they now had a powerful incentive not to do so. At any time during the loan period, they could redeem the loan at the daily posted market price in their county, even when this price was well below the amount of the loan. This gain was called a market loan payment. Those farmers who chose not to take out loans could request the same benefit (called a loan deficiency payment) at the time of low posted prices. With this system, plus very effective export efforts, the CCC was able to eliminate almost all stored surpluses by 1990.[6]

In response to intense lobbying by major environmental organizations, and in the wake of increased erosion during the expansion of grain farming in the booming 1970s, the 1985 bill included new and very strict compliance rules for its conservation grants. In fact, this farm bill will always be remembered as the "Environmental Act." For example, farmers who owned highly erodible land, such as in the Palouse area of eastern Washington, had to submit a plan to cure the problem and carry it out within five years in order to retain any payments. Farmers who planted crops on wetlands forfeited their eligibility for loans and deficiency payments. Congress also revived the Conservation Reserve program. But the high level of payments (up to $26 billion in 1986) created public concern about a farm policy that had somehow gone badly awry. Fortunately, farm prices began a gradual recovery in 1986, as surpluses vanished. One other small program helped. A group of cooperating farmers could form a Farm-Owned Grain Reserve, which earned them lower loan interest rates and payments for local storage of retained commodities. For the most part, the 1985 farm program continued until 1996, but early in the Clinton administration, budgetary restraints led to one significant reduction—farmers could gain payments on only 85 percent of their base acres. By then, international issues were impinging on farm legislation and marketing policies.

INTERNATIONAL AGREEMENTS AND THE FEDERAL AGRICULTURAL IMPROVEMENT AND REFORM ACT

Despite high agricultural price supports and payments in the United States (up to $20,000 a year for each full-time farmer in 1985), import quotas or export subsidies in the European Union were even more market distorting on a per capita basis, and up to four times more so in Japan. Such policies impeded U.S. exports, raised domestic food prices, and practically excluded farmers in poor countries from Western markets. In the international trade negotiations leading to the General Agreement on Tariffs and Trade (GATT), the United States pushed for a much freer market, even though its own policies, which it defended as necessarily defensive in nature, clearly violated such principles. By the next omnibus farm bill in 1990, the GATT negotiations were under way, but they would not lead to a final agreement until 1994, which completed the so-called Uruguay Round that had started in 1986. GATT created the World Trade Organization (WTO) to monitor and enforce these agreements.

The stated goal of the long and difficult negotiations leading to the 1994 agreement was "a fair and market-oriented agricultural trading system." The agreement moved toward such a market system, but not very far because of compromises to gain European Union agreement and a large number of exemptions from the most demanding rules. The main goal was a drastic reduction in market-distorting policies. Thus, the complex rules tried to distinguish between benign farm policies with little impact on international trade and those that clearly distorted it. Although the United States, which then led the world in the production of corn, wheat, and soybeans, was the leading exponent of freer trade, in the actual implementation of the agreement it resorted to subterfuge to evade the intended effect of the new policies.

At first, the agreement included two categories of policies—those that significantly distorted trade ("amber box" policies) and those that had little or no impact ("green box" policies). To achieve final agreement, the negotiators had to accept some seemingly amber policies followed in the European Union and the United States, such as production-limiting controls on acreage or yields or on numbers of livestock, if those applied to no more than 85 percent of base production. This compromise led to a special "blue box" category. The amber box of trade-distorting policies encompassed most existing U.S. farm policies, including direct payments to producers based on units of production and subsidized loans or crop insurance.

Green policies—those with a negligible (an imprecise word) impact on production and trade—included support for research and education, pest and disease control, marketing promotion, public infrastructure, and even public storage of crops if collected on behalf of food security and purchased at market prices. None of these activities could involve direct payments to producers. Some payments to individual producers were acceptable, such as direct income support (welfare) when not tied to production or prices, safety net programs that did not involve market issues, natural disaster assistance, land retirement or other structural changes not tied to production, payments for environmental or conservation goals, and special assistance to producers in disadvantaged regions. These allowable policies were subject to interpretation and, in some cases, such as disaster aid, allowed countries like the United States to finesse many of the purposes of the overall agreement.

With the possible exception of New Zealand, which had almost no agricultural subsidies, none of the negotiating countries was willing to exclude all amber policies. Instead, they only tried to limit them. The countries agreed to reduce amber policies by 20 percent from a base period, usually 1986–1988, and phase in these changes over a six-year period. By design, these were years of high price support with a high starting base. At the end of the six years, in 2000, the United States could spend just over $19 billion on amber policies without violating the agreement. This quota had to be apportioned on a crop-by-crop basis, with non-crop-specific subsidies aggregated into one class. Even here, the agreement made one exception: if the subsidy for a crop totaled less than 5 percent of the crop's market value, it did not count against the national quota. As one would expect, the countries that were party to this agreement have, by their own interpretation, stayed within their amber box quotas, although many, including the United States, have faced challenges on the criteria used to calculate what fit into that box.[7]

The 1994 international agreement helped shape a new farm bill in 1996, and all subsequent farm bills have included escape clauses in case total amber spending exceeded the ceiling. No change in policy since the Agricultural Adjustment Act of 1933 seemed nearly as important as the 1996 Federal Agricultural Improvement and Reform Act (FAIR). For Title I on commodity programs, FAIR seemed revolutionary. For all commodities except tobacco, it eliminated all the old touchstones of farm policy—no more acreage controls and, because crop loans would now be at or below market prices, no more distorting price supports and no more stored surpluses (which had all but ended even before 1996) beyond congressionally mandated reserves. Even the title, from the perspective of long-term critics of farm subsidies, suggested a new era of fairness, while Republican legislators dubbed it a "freedom to farm" bill. The complicated old system was dead, and few regretted its passage. But at the same time, farmers were aware that they had lost a safety net. They made Congress aware of their new insecurity, and Congress responded with some wonderful but short-term medicine.[8]

GATT forced the United States, whenever it threatened to surpass the $19 billion ceiling, to decouple any aid program from annual production. The 1996 farm bill did much of this decoupling. Thus, to help farmers make the transition from production-based payments, which in some years exceeded 20 percent of their net income, Congress provided "production flexibility contract payments" over the next seven years. These had no direct ties to present production, for the amount of each payment was based on base acreage and past production records. This was a ruse to save the base and to avoid amber box involvement. Farmers could collect the payments even if they stopped growing crops; likewise, a new purchaser of a farm with base acreage could collect payments, even if the purchaser did not continue to grow crops (a scandalous provision removed in most proposed versions of a 2007 farm bill). The payments were not large on a unit basis (for example, 28 cents per bushel for corn). But for very large farmers, this could be a significant amount of money. Though intended as an income-supporting payment, it provided the most funds to the ablest and most successful large farmers because, as always, they found ways around the payment limits under the new program. One could also argue that these payments influenced farmers' choices—in some cases, helping inefficient farmers remain in agriculture—and thus were market distorting.

The marketing loan program continued, but in a manner that seemed devised to gain price stability rather than provide a subsidy to farmers. The loan rate was capped at 1995 levels for corn and wheat, a good year for farm prices, but it was nonetheless expected to parallel market prices. It quickly added to the amber box quota of $19 billion, since farm prices fell during the last years of the twentieth century. Thus, the loan rate for most crops was soon well above the market price, guaranteeing the loan rate for farmers who forfeited their loans at the end of the contract period (usually nine months). By 2000, this loan gain reached $8 billion. In addition, if prices sagged, farmers could redeem their loans at posted county price levels and gain an added subsidy. Farmers could also forgo loans and gain a loan deficiency payment equal to the difference between the loan rate and the local price; this tactic was followed by a majority of farmers, with costs of more than $6 billion by 1999. Even so, gains from this loan program were not large enough to violate the $19 billion quota. Even when joined with other subsidies, such as those tied to highly subsidized crop insurance, they did not risk WTO limits.[9]

Because of falling prices during the late 1990s, the transition payments and the continued loan program did not maintain farm incomes at 1995 levels. Farmers were freer but, as it turned out, much less secure. This put pressure on Congress to come to their rescue, and as always, it did so. For the 1999 farm year, it began appropriating what it called "emergency market loss payments." These ad hoc payments grew from $3 billion in 1999 to $11 billion in 2000, when federal subsidies to farmers reached an all-time record, at least in current dollars (more than $25 billion, or 47 percent of farmers' net income). Prices improved in 2001, and these supplemental payments fell, but the precedent had been established. By the time of the 2002 farm bill, these payments came close to being a new entitlement for farmers. By any fair evaluation, these payments should count as part of the amber box.

Beginning in the 1970s, Congress had become increasingly generous in passing emergency disaster relief for farmers. At first, most such assistance consisted of subsidized loans, but in almost every year after 1988, Congress used ad hoc bills to offer direct payments to farmers in counties that suffered major crop losses from flood, drought, wind, or disease. The minimum loss required to qualify was usually 35 percent of the normal yield, and the payments were usually 65 percent of the estimated market price of the afflicted crop for farmers who had bought into the heavily subsidized crop insurance program. Those with insured crops could recoup all or more than their normal income. The rate of relief dropped to 60 percent for eligible farmers who had rejected crop insurance. In some years the receipt of such a disaster payment required a farmer to buy crop insurance in the next two years. Emergency aid for livestock owners usually required a 40 percent loss of available grazing for at least three months. In the 2000 crop year, such disaster payments totaled $1.8 billion. They soared in succeeding years as Congress lowered the threshold for disaster designation. The disaster area always included an entire county, which often meant that some farmers eligible for disaster payments had harvested normal crops. In 2003 all 256 Texas counties were eligible for emergency livestock assistance.[10]

The 2002 Farm Bill and Beyond

To the despair of many critics of federal commodity policies, FAIR had failed. Nothing much had changed. Payments to farmers rose rather than fell, and with acreage reduction a memory, there seemed no way to bring production into line with demand and thus no way for taxpayers to es-

'cape what had become a generous welfare state for the roughly 300,000 farmers who produced 89 percent of total farm output. The 2002 farm bill simply reinforced the new security system by converting the supplemental or ad hoc payments into counter-cyclical payments that largely resumed the target prices and deficiency payments in effect through 1995. Based on another name change in 1995, all the commodity programs remained under the direction of the Farm Service Agency, the most recent incarnation of the old Agricultural Adjustment Administration of 1933.

The 2002 bill expanded the number of crops eligible for crop loans, with peanuts added to at least sixteen other crops. Of the major storable crops, it excluded only tobacco. Congress set the loan rate (in pounds or bushels) for each crop. Farmers could forfeit the loan at the end of the contract period. Alternatively, they could redeem the loan at any time at the posted daily county price for grains and soybeans and at the weekly world price for cotton and rice, a tempting option when prices fell well below the loan rate. Gains from this loan program clearly fit in the amber box.[11]

The transitional payments of 1996, scheduled for a seven-year period, became a five-year guaranteed subsidy in the 2002 bill and applied to even more crops (about ten in all). The legislation set the payment schedule for productive units in each crop (corn, for example, at 28 cents a bushel). These "direct payments" covered only 85 percent of base acreage (this provision fit the blue box terms of the WTO). One farmer might draw payments for several different crops, but the upper limit was set at $40,000 per person, which was effectively doubled if one had a spouse or farmed on three different farms, or "entities." The only other stipulation was that farmers comply with conservation or wetland programs. Because these payments were based on past production and in no way limited present crop choices, the United States gave them a green box classification.

The new part of the triad of subsidies was counter-cyclical income support payments, which became a fixed-rate replacement for the various supplemental payments Congress had provided annually from 1998 to 2001. These counter-cyclical payments were based on the same 85 percent of base as the direct payments, with the unit amount set by legislation. They applied only when farm prices fell below a target price set, at a shifting annual rate, by the bill itself. If crop loan gains and direct payments raised returns from a given crop above the target price, then the farmer would not be eligible for counter-cyclical payments. The maximum payment was $60,000, but with the same escape clause for those with spouses or three farming entities. Wealthy farm owners, such as successful businessmen or physicians, whose adjusted gross income exceeded $2.5 million were ineligible for any payments unless two-thirds of that income was from farming. Because the amount of the payment was determined by past production levels, the United States claimed green box status for these payments, but other countries challenged this classification. Brazil won an appeal to the WTO about American cotton subsidies, and Mexico and Canada have pending appeals about rice and corn payments.

Fortunately for farmers and taxpayers, few crop loans or counter-cyclical payments were needed after 2001. Most farm prices increased in these years, fueled in part by the ethanol and biofuel boom and, more recently, by a worldwide shortage of wheat stores. Most farmers, at least in the commodity areas, are happy with their generous support programs. They now have the best of all worlds—the right to produce all they want, and a government security blanket to protect their incomes.

The 2002 farm bill was due to expire in September 2007. Thus, throughout 2007 Congress debated a new bill that would cost about $300 billion over the next five years. The House completed a timely bill in July, but a conflicted Senate only passed its version in October. This delay required the first of seven eventual extensions of the 2002 bill, until June 6, 2008. A conference committee first convened in April 2008, and only came close to a final agreement on May 5. The original Senate bill had added a new $5 billion permanent disaster relief trust fund and offered more than $2.5 billion in new tax exemptions or credits to farmers or agribusinesses (essentially earmarks pushed by individual senators, with the most publicized being tax reductions for racehorse owners). The House resisted such costs for a bill already well above the 2002 level, but in conference succeeded only in reducing the relief fund to $3.8 billion and the tax breaks to $1.8 billion.

The economic and political environment shifted radically during the extended debates. A surge in farm prices by early 2008 led to a global food crisis and a rapid increase in food prices in the United States. Suddenly, the existing subsidies for American farmers and the diversion of one-fourth of corn production into ethanol seemed irresponsible, or even immoral. Yet, the final conference bill made only minor changes in the commodity programs. It lowered the annual direct payments ($5.2 billion) by a token 2 percent. It slightly raised some loan rates or counter-cyclical target prices. In May 2008, most farm prices (corn at more than $6 and wheat at about $10 per bushel) were at double these levels, meaning no governmental costs in the near future. But farmers wanted the security these programs would provide in future years if the boom, like the one in the mid-seventies, led to another bust. The bill added research and marketing subsidies for fruit and vegetable growers (a first), expanded several conservation programs, and added subsidies for the development of cellulosic ethanol while lowering the subsidy for corn ethanol from 51 to 45 cents a gallon. Most costly was the addition of more than $10 billion for nutrition programs, primarily for a badly needed reform of the food stamp program. This meant that two-thirds of the cost of the farm bill would be for food aid.

Through all the deliberations, one issue divided the Bush administration and Congress—payment limits. The administration wanted to exclude from all commodity payments any individual farm operator whose adjusted gross income exceeded $200,000. Since the existing income cap was $2.5 million, this was a revolutionary proposal, one that could have cut the cost of the commodity programs by up to half. The administration argued that American taxpayers should not have to subsidize individuals with income levels in the top 2 percent, including many affluent physicians, lawyers, and businessmen who owned farms. As one might expect, large commercial farmers, those with the most political clout, were dismayed by such a restriction. The conference committee did not adopt limits acceptable to President Bush, who promised to veto what he called a "bloated bill." But both the House and the Senate passed the bill by May 15 with such overwhelming majorities as to assure an override of any presidential veto. Thus, what is now the farm bill of 2008 will control agricultural polices through the crop year of 2012.

Noncommodity Programs

Commodity price supports are the most visible, most direct, and, except for food aid, most expensive facets of American agricultural policy. But commodity programs from the 1930s on have always existed beside other means of supporting higher incomes for farmers. Most important in the two decades after World War II were various efforts to expand exports. The story is complex and is directly related to U.S. foreign policy. At the end of World War II, various agencies distributed food to needy European countries. The Marshall Plan required participating European countries to use their own currencies to purchase food from the United States, with many of these credits returned to Europe for development projects. Various loans and credits facilitated the sale of food abroad. These programs, many tied to foreign aid, came together in 1954 with the Trade Development and Assistance Act, or what is generally known as Public Law (PL) 480. In the early years, most exports involved gifts or loans, or sales to countries in their own currencies. Some of these would be returned as a type of development aid. A few barter transactions also took place. But in 1966 a new Food for Peace program normalized most of this trade. Subsidies to American exporters allowed them to buy, at reduced rates, higher-priced American farm products. By 1973, PL 480 exports totaled more than $23 billion. Today, more than $1 billion a year (small in comparison to other programs) is expended under PL 480 or later amendments to it through more than ten different programs, some targeted at agricultural development in poor countries. All these programs help maintain demand for American farm products, as do ad hoc appropriations for food relief in the case of emergencies or famines in various parts of the world.

Since the New Deal, the federal government has always offered farmers subsidies on behalf of various conservation projects. Today, more than a dozen different programs are in place, but two are most important. One is a new Conservation Security Program. It provides enhancement grants for a wide range of conservation practices in a growing number of major watersheds. Farmers submit proposals for projects involving erosion control, water management, forest improvement, and watershed protection. This is an updated version of the conservation payments that earlier supported farm ponds and terraces. More daring is the Environmental Quality Incentive Program, or EQIP. It was first established in 1997 to help farmers deal with requirements under clean air and water acts, wetlands legislation, and the Endangered Species Acts. It makes incentive and project grants to help farmers reduce nonpoint pollution; improve methods of handling farm waste, including manure; reduce emissions from farm machinery; reduce soil erosion; and protect wildlife habitats. It is particularly useful to farmers committed to a more sustainable form of agriculture. These grants may encompass new rotation patterns, experiments with low- or no-pesticide farming, energy-saving techniques, recycling strategies, and wildlife protection. As one might expect, small farms or those with limited net income are most likely to avail themselves of these two programs.

Land retirement schemes began in the 1930s as a means of reducing surpluses and achieving certain conservation goals. After World War II, several conservation programs continued, providing federal subsidies for erosion control, soil building, and the development of wildlife habitats. Land retirement as a crop reduction program revived only in 1956, with the short-lived Soil Bank and Conservation Reserve program (discussed earlier). The present Conservation Reserve was born in the 1985 farm bill, at the climax of the worst agricultural depression in the postwar era. It had three goals: cut the then-growing surplus, remove highly erodible land from production, and improve the environment. It offered farmers an opportunity to retire erodible land from production for ten years. The acres leased to the government had to reduce, acre for acre, the base acreage for program crops (on which various payments were based), thus curtailing production and income. In the original plan, farmers were to negotiate the rent received for retired land, but the congressionally mandated maximum rent soon became an almost universal standard. The average rent per acre was approximately $50 in 2007, thus far below the gross income (but not necessarily the net income) from active farming in most parts of the country. For certain people, including some retired farmers, it offered a way to hold on to land and gain at least enough income to pay taxes. Farmers were required to create protective cover, sometimes with financial assistance, mow the grassland annually, but not harvest the hay except in certain drought situations. Supplemental reserves of farmable wetlands and enhanced programs that led to state-federal cooperation in specific environmental projects have been added to the program, which was originally intended to retire at least 40 million acres, or just over 4 percent of cultivated land in America. It has not reached that goal, but in 2007 the reserve contained nearly 37 million acres. Its role in curtailing food production has been slight, but the individual reserves (renewable at the end of ten years) have created wildlife habitats, watershed protection, erosion controls, and new forestland. Despite the qualification that the lands be highly erodible, I am familiar with excellent, though hilly, farmland in the reserve. Should circumstances require more food production in the United States, the major part of this reserve awaits its reentry into farming.[12]

The present 51-cents-per-gallon subsidy for ethanol producers is an indirect farm subsidy. It has increased the market for corn and raised its price, to the benefit of corn farmers. Indirectly, it has raised the prices of other grains and soybeans, which have a smaller but growing market for biodiesel fuels. The ethanol boom has already raised the cost of foodstuffs, and if present projections hold, it will continue to do so in the future, when the present 20 percent of corn that goes into ethanol will skyrocket. It remains unclear how much energy the use of ethanol saves, but at least indirectly, it probably reduces greenhouse emissions (how much depends on the fossil fuels that go into the production of ethanol).

When biomass is used as a fuel, the carbon dioxide (CO_2) emitted is that just absorbed from the air by growing plants. In this sense, it makes no net addition to atmospheric CO_2. Since only the grain is used in ethanol, the CO_2 in cornstalks remains sequestered until they decay or, possibly in the near future, they are used in the production of cellulosic ethanol.

In monetary terms, the largest Department of Agriculture program aimed at the creation of demand for farm goods involves domestic food aid. In 2008 all food programs were scheduled to cost $57 billion, or 64 percent of the total budget of the Department of Agriculture. The first federal food aid programs began in the Great Depression and consisted of a wide distribution of surplus commodities. In 1933 Congress formed the Federal Surplus Relief Corporation, which worked in conjunction with the state-based Federal Emergency Relief Administration. In 1935 a new Federal Surplus Commodities Corporation became an enduring agency for food purchases and disposal. As part of an amendment to the Agricultural Adjustment Act of 1933, section 32 provided that one-third of all customs receipts be used in support of agriculture—to facilitate exports, encourage domestic consumption, and reestablish farmers' purchasing power. This legislation underwrote much, but not all, food relief until the early 1970s, when the food stamp program became fully established. But several food distribution programs, some involving section 32 funds, have continued to the present. The government has explored every possible way to get rid of surplus food, including giving it to charitable institutions, county welfare agencies, schools, and needy families. In the postwar years, the variety of foods could be limited to only eight or ten in excess supply. But after 1954 the government expanded the inventory and sought a balanced, nutritional diet. With the world food shortage in the early 1970s, the Department of Agriculture had to buy some foods for the program in the open market, because it had no surpluses. The program sometimes distributed more than a billion pounds of food a year to almost 3 million institutional or family recipients, but the cost, at least by contemporary standards, was modest—up to only $333 million a year.

In 1946 Congress gave enduring form to what had been an on-and-off distribution of food to some schools during the Depression. The National School Lunch Act had a dual purpose—to safeguard the health of children and to encourage the domestic consumption of nutritious agricultural commodities. The act offered federal matching funds to states to support lunch programs, but it also provided for the direct distribution of food to schools. Participating schools had to serve lunches meeting minimal nutritional standards, provide such meals for free or at reduced cost to poor children, and use surplus or donated commodities to the extent possible. Eligibility requirements for matching funds were lower for states with below-average per capita incomes. By 1973, more than 86 percent of schools participated in the program, almost half of all children received school lunches, and more than a third of these children qualified for free lunches. In 1967 the program expanded to include school breakfasts for disadvantaged children. By 1995, the federal cost exceeded $6 billion a year; by 2007, it cost $8.2 billion, along with $5.7 billion for other child nutritional programs. By then, it served 31.5 million children in 101,000 schools.

A smaller program began with the Child Nutrition Act of 1967. It gave grants to the states for a Women, Infants, and Children (WIC) program that now costs $5.5 billion a year. It supports food aid for children up to five years old, provides nutritional assistance to mothers, and even gives referrals for necessary health care. It is basically a welfare program, with only a limited impact on overall food demand.

By far, the most costly and important domestic food aid is the food stamp program, which cost more than $37 billion in 2007. Food stamps were used to provide food aid for a brief period in the late 1930s and early years of World War II, or until food shortages replaced surpluses. The Kennedy administration reintroduced pilot programs in 1961. As part of the War on Poverty, Congress enacted the Food Stamp Act in 1964, with major amendments in 1977 and 2008. The first listed purpose of the act was not welfare but the expansion of demand for surplus commodities. The Department of Agriculture administers the program, which in almost all years costs more than all price support payments combined. From the beginning, families with incomes less than 130 percent of the official poverty level, and with liquid assets of no more than $2,000 were eligible for food stamps. Eligible families paid variable amounts for the stamps, based on family size and need, but always less than their monetary value. Recipients could use them in grocery stores for most food items, but not for alcohol or tobacco or, in the early years, for imported foods, particularly imported meats (this highlights the surplus-reducing purpose of the stamps). Today, the enduring label food stamps is anachronistic, for the system has fully converted to electronic transfers, using plastic debit cards comparable to bank debit cards. The debit cards have eliminated a great deal of fraud, for some recipients used to sell their stamps (some still sell the food). It is impossible to determine how much food stamps add to the overall demand for food and thus how much they add to the income of farmers, since the families receiving this assistance would have found other ways to get food, including charity. Unfortunately, despite several educational efforts, it is not possible to get all food stamp recipients to buy the most nutritious or the most economical foods (present legislation includes funds for nutritional education). Even if it does little to increase the demand for crops, food aid has certainly benefited farmers in a political sense. In the debates on the 2007 farm bill, increases in food aid helped deflect opposition to the commodity programs.[13]

These various methods of supporting the demand for farm products often run counter to other governmental policies. This includes the ever-expanding research agenda of the Department of Agriculture and of the land-grant universities that carry out most of the research. Research helps increase production, as do the highly subsidized irrigation projects, largely in the West, and the various credit programs, including some that encourage young people to enter agriculture. Rural development programs, most operated by the Department of Agriculture, improve the quality of farm life, although only a small percentage of rural Americans live on farms today (less than 10 percent). About half of farm families live within statistical metropolitan areas, and thus close to larger cities. In my state the leading farm county is within the metropolitan area that centers on Nashville. Many of the counties that have received the most funds for rural development are in parts of the Great Plains or the South that are far from urban complexes, but even here most residents are not farmers. In fact, one count in 1994 revealed only 556 farming counties in the United States, or counties in which 20 percent of earned income came from agriculture (and even in these counties, other areas of employment usually exceeded agriculture).

This survey of federal support programs for farmers only scratches the surface of the huge Department of Agriculture and the hundreds of different programs supported by a budget approaching $100 billion a year. One need only read the proposed farm bill of 2007 to appreciate the complexity of farm programs or the density of the bureaucratic prose that describes them.

7. Farming in the Twenty-first Century

Status and Challenges

It is impossible to calculate the exact number of farms and farmers presently in the United States. Since 1974 the Department of Agriculture has defined a farm not by acreage but by the volume of sales. The production of agricultural goods that sell for $1,000, or would normally sell for that amount, qualifies an operation as a farm. In 1999 it added a few categories that do not meet this criterion: horse farms with five or more horses, even if sales did not amount to $1,000, and farms producing maple syrup or short-rotation woody crops. A farm census is taken every five years, and in 1997 the responsibility for this survey moved from the Census Bureau to the National Agricultural Statistics Service in the Department of Agriculture. The last farm census for which statistics are available was in 2002 and involved not only the survey of farmers but also sophisticated sampling tests to make up for the inevitable undercount. This revealed a total of 2,128,982 farms. Note that in any given year, some small farms will move out of this class because of low sales, while others will cross the $1,000 threshold.[1]

PROFILE OF CONTEMPORARY FARMS

Agriculture is, by far, our least diverse economic sector in terms of race, ethnicity, and gender. More than 97 percent of principal farm operators are white, and just over 90 percent are male, although women make up a much larger share of nonprincipal or secondary operators. African Americans, once so critical to southern agriculture, have almost completely deserted farming. Only 29,090 are principal operators, meaning either owners or tenants. Farm operators are older than individuals in most occupations: principal operators average 55.3 years of age, with an older average in the small farm categories. The trend is for the average age of farmers to increase in the next few decades, in part because entry into farming is so difficult for young people. The average number of people in each farm household is just over two, or close to the national average. Finally—and I must emphasize this point to skeptical readers—almost all farms are owned by families (this includes extended families, in the sense that a father and son, with two households, may jointly own a farm), not by corporations that are not owned by members of a family. In 2002 only 7,661 farms were primarily owned (more than 50 percent of the stock) by nonfamily investors, and these accounted for only 9,319,000 acres out of a total of 938 million acres, or only 1 percent of the total. It is true that these corporate farms are much larger than average, as are the more numerous (66,667) incorporated farms owned by families.[2] (See table 1.)

But family ownership does not necessarily correlate with what most people mean when they refer to family farms. Of course, no one definition of a family farm would please everyone, particularly among those who farm and believe themselves to be "family" farmers. Some traditional criteria associated with the much revered family farmer fit only a small minority of those who operate American farms. The following are among the most prevalent criteria: that a family owns, lives on, and works on a farm; gains most of its livelihood from farming; and, perhaps most critical, makes all the important decisions about the farming operation. This managerial independence has supported most images, or myths, about farmers as responsible citizens. According to these criteria, most officially listed American farms are not family farms. On a majority of these farms, the operator does not gain even half of his or her income from farming. Nearly a fourth do not live on farms. And even among full-time farm operators, some of the largest and wealthiest farmers do not have full managerial control. This includes most of the factory-like chicken and hog farms that are under contracts to large corporations. Finally, many very large family-owned and -operated farms, such as a California dairy farm with more than 1,500 cows and a dozen or more employees, seem more like urban business enterprises than traditional farms.

Table 1. Historical Highlights: 2002 and Earlier Census Years (2002 Farm Census)

All farms		2002	1997	Not adjusted for coverage					
				1997	1992	1987	1982	1978	1974
Farms (number)		2,128,982	2,215,876	1,911,859	1,925,300	2,087,759	2,240,976	2,257,775	2,314,013
Land in farms (acres)		938,279,056	954,752,502	931,795,255	945,531,506	964,470,625	986,796,579	1,014,777,234	1,017,030,357
Average size of farm (acres)		441	431	487	491	462	440	449	440
Estimated market value of land and buildings [1]									
Average per farm (dollars)		537,833	416,007	449,748	357,056	289,387	345,869	279,672	147,838
Average per acre (dollars)		1,213	967	933	727	627	784	619	336
Estimated market value of all machinery and farm equipment [1] ($1,000)		136,624,880	119,302,923	110,256,802	93,316,496	85,801,360	93,662,947	77,600,689	48,402,624
Average per farm (dollars)		66,570	53,861	57,678	48,605	41,227	41,919	34,471	22,303
Farms by size									
1 to 9 acres		179,346	205,390	153,515	166,496	183,257	187,665	151,233	128,254
10 to 49 acres		563,772	530,902	410,833	387,711	412,437	449,252	391,554	379,543
50 to 179 acres		658,705	694,489	592,972	584,146	644,849	711,652	759,047	827,884
180 to 499 acres		388,617	428,215	402,769	427,648	478,294	526,510	581,631	616,098
500 to 999 acres		161,552	179,447	175,690	186,387	200,058	203,925	213,209	207,297
1,000 to 1,999 acres		99,020	103,007	101,468	101,923	102,078	97,395	97,800	92,712
2,000 acres or more		77,970	74,426	74,612	70,989	66,786	64,577	63,301	62,225
Total cropland	farms	1,751,450	1,857,239	1,661,395	1,697,137	1,848,574	2,010,609	2,081,604	2,157,511
	acres	434,164,946	445,324,765	431,144,896	435,365,878	443,318,233	445,362,028	453,874,133	440,039,087

(continued)

Table 1. *(continued)*

All farms		2002	1997	Not adjusted for coverage					
				1997	1992	1987	1982	1978	1974
Harvested cropland	farms	1,362,608	1,545,681	1,410,606	1,491,786	1,643,633	1,809,756	1,904,602	1,954,700
	acres	302,697,252	318,937,401	309,395,475	295,936,976	282,223,880	326,306,462	317,145,955	303,001,943
Irrigated land	farms	299,583	308,818	279,442	279,357	291,628	278,277	280,779	236,733
	acres	55,311,236	56,289,172	55,058,128	49,404,030	46,386,201	49,002,433	50,349,906	41,243,023
Market value of agricultural products sold ($1,000) [2]		200,646,355	201,379,812	196,864,649	162,608,334	136,048,516	131,900,223	107,073,458	81,526,126
Average per farm (dollars)		94,245	90,880	102,970	84,459	65,165	58,858	47,424	35,231
Crops, including nursery and greenhouse crops ($1,000)		95,151,954	100,668,794	98,055,656	75,228,256	58,931,085	62,256,087	48,203,200	41,790,365
Livestock, poultry, and their products ($1,000)		105,494,401	100,711,018	98,808,993	87,380,078	77,117,431	69,644,136	58,870,258	39,503,850
Farms by value of sales [3]									
Less than $2,500		826,558	693,026	496,514	422,767	490,296	536,327	460,535	649,448
$2,500 to $4,999		213,326	265,667	228,477	231,867	262,918	278,208	300,699	257,263
$5,000 to $9,999		223,168	267,575	237,975	251,883	274,972	281,802	314,088	296,373
$10,000 to $24,999		256,157	293,639	274,040	301,804	326,166	340,254	394,876	(NA)
$25,000 to $49,999		157,906	179,629	170,705	195,354	219,636	248,828	300,515	(NA)
$50,000 to $99,999		140,479	163,510	158,160	187,760	218,050	251,501	263,092	(NA)
$100,000 to $499,999		240,746	282,422	277,194	286,951	263,698	274,580	203,695	141,187
$500,000 or more		70,642	70,408	68,794	46,914	32,023	27,800	17,973	11,412

(continued)

Table 1. (continued)

All farms	2002	1997	Not adjusted for coverage					
			1997	1992	1987	1982	1978	1974
Farms by type of organization								
Family or individual	1,909,598	1,922,590	1,643,424	1,653,491	1,809,324	1,945,639	1,965,860	(NA)
Partnership	129,593	185,607	169,462	186,806	199,559	223,274	232,538	(NA)
Corporation	73,752	90,432	84,002	72,567	66,969	59,792	50,231	(NA)
Other--cooperative, estate or trust, institutional, etc.	16,039	17,247	14,971	12,436	11,907	12,271	9,146	(NA)
Principal operator by days of work off farm [4,5]								
None	962,200	832,585	755,254	801,881	844,476	861,798	942,803	829,843
Any	1,166,782	1,254,537	1,042,158	992,773	1,115,560	1,187,374	1,203,286	1,011,476
200 days or more	832,348	870,945	709,279	665,570	737,206	774,844	770,045	657,971
Principal operator by primary occupation [5]								
Farming	1,224,246	1,044,388	961,560	1,053,150	1,138,179	1,234,787	1,269,305	1,427,368
Other	904,736	1,171,488	950,299	872,150	949,580	1,006,189	988,470	851,902
Average age of principal operator [5] (years)	55.3	54.0	54.3	53.3	52.0	50.5	50.3	51.7
Total farm production expenses [1] ($1,000)	173,199,216	157,752,357	150,590,993	130,779,261	108,138,053	(NA)	(NA)	61,007,649

1. Data are based on a sample of farms.
2. Data for 1974 include the value of forest products sold.
3. Data for 1982 and prior years exclude abnormal farms.
4. Data for 1997 and prior years do not include imputation for item nonresponse.
5. Data for 1974 apply only to individual or family operations (sole proprietorships) and partnerships.
(NA), not available

Narrowly specialized, large-scale farming departs widely from most images of the traditional family farm. In what sense does a 10,000-acre wheat farm in Kansas fit any definition of a family farm, when machines do all the planting and harvesting and the operator does not even live on the farm? Or what about a family-owned chicken farm with 20,000 broilers or 2,000 laying hens all housed in one-square-foot stacked cages in a large coop or barn, with no direct contact between the owner and the chickens? Or the 18,000 hog farmers who produce more than 90 percent of the total product, or the 11,000 beef cattle owners with annual sales over $1 million who market just over half our beef?

Such huge operations do not match nostalgic memories of family farms, where livestock played an indispensable role. Then, the ties between animals and family members were close, in a sense personal. Cows had names. Chickens flocked around the wife who scattered the cracked corn. Against all expert advice, local farmers in my boyhood community insisted on keeping roosters for the pleasure of the hens and thus sold fertilized eggs and, in some cases, would eat no other. After a year, hogs that had been lovingly fed and cared for faced a necessary slaughter. But in my community, no one would do what was best for the meat—stun a hog, hang it by its hind legs, and cut the jugular artery in the throat to let the beating heart effectively remove the blood. That seemed too cruel. Instead, they shot the hog in the head to minimize its suffering, and only then slit the throat. In my home, my sister and I cried when we heard the shots, and my father could not bring himself to do the shooting. He hired someone else to do it. It is this type of farming on a small, human scale that many people identify with the true "family farm." They see it as more of a way of life than a means to gain profits, or as a form of artful engagement rather than just a job.

All states have farmers, although very few are in Alaska. In the last two decades the former farm states, at least as most people conceived of them, have slipped in their share of income. This is true in the Midwest, with its dominance in corn, wheat, and hogs, and in the South, with its lead in cotton and tobacco (both now declining crops). In 2007, as ranked by total farm income, California and Texas easily led all other states and will continue to do so. But perhaps surprisingly, North Carolina ranked third, with its leap ahead based largely on factory-like hog production. Iowa, the traditional leader in corn, is now fourth in total farm income. The other top ten, in rank order, are Florida, Minnesota, Nebraska, Georgia, Kansas, and Kentucky. Little Delaware leads all states in farm income per capita and per acre, based largely on its poultry factories, which require few inputs of either land or labor. Poultry has also boosted Arkansas to thirteenth. The shifting patterns, largely influenced by the efficiency of poultry and hog farms, have left former leaders far behind. For example, Illinois now ranks twenty-seventh, despite its traditional first or second rank in corn, and Wisconsin, the Dairy State, is now fifteenth in overall income and trails California even in dairying.

The relative importance and value of different crops have also shifted, though not radically. Such traditional leaders as corn and wheat are still near the top, but they have been joined by soybeans (see figure 11). In fact, in 2002 the number of acres planted in soybeans (more than 72 million) was slightly larger than those planted in corn for grain (68 million). If one counts corn planted for silage, it remained the leader. Since 2002, the ethanol boom has pushed corn well above soybeans, or to 92 million acres in 2007. Hay crops of many types (64 million acres) came in third, ahead of wheat (45 million acres). In 1950 very few farmers knew anything about soybeans, let alone grew them. At the same time, cotton was one of the five leading crops. No more. Cotton is down to 12 million acres, and tobacco is down to only 428,000 acres and, I hope, still declining. Sorghum, as a feed grain, has exceeded the acreage of oats and barley combined. Orchard crops account for more than 5 million acres, rice just over 3 million, and vegetable crops 3.5 million. No other crops require more than 2 million acres. But note that acreage does not

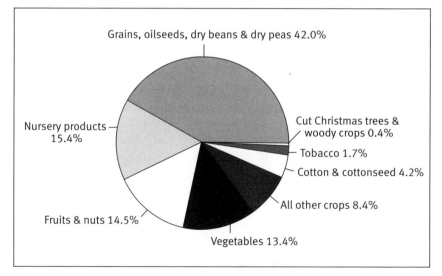

Figure 11. Value of crops sold: 2002. Total = $95,151,954,000.
(2002 Farm Census)

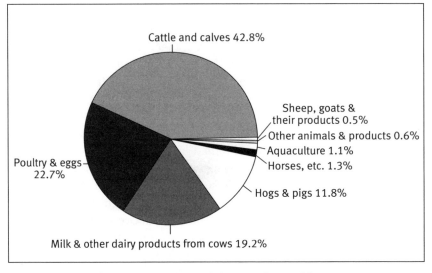

Figure 12. Value of livestock, poultry, and their products sold: 2002. Total = $105,494,401,000. (2002 Farm Census)

always correlate with value, especially in the case of vegetables, fruits, and nursery stock, which collectively have almost the same value as corn, wheat, and soybeans combined.[3] Among livestock, what is most significant is the rapid rise of poultry (see figure 12).

Farm Labor

Who does the work on farms? How many farmers are there in the United States? It is not easy to give a firm answer, in large part because so many farm operators have off-farm employment and often devote only a few hours a week to managing or working on a farm. It is also complicated by the increasing number of temporary and, in some cases, migratory farmworkers. A century ago it was more simple. In 1910 around 10 million family members lived and worked on farms (this number excludes young children and, in most cases, housewives). About 6 million of these were farm operators, either owners or cash or share tenants, and 4 million were other adult family members who worked full time. In addition, these 6 million operators employed about 3.5 million hired hands, most of whom lived on the farms where they worked. They made up just over one-fourth of all farmworkers. Roughly the same ratio prevailed until the climax of the agricultural revolution in 1970. Since then the total number of wage workers has increased, now accounting for 30 to 35 percent of all farm employment. But the structure of wage labor has changed dramatically. Few live-in hired hands remain. Less than half of today's farmworkers are employed on a full-time basis. And the percentage of farm costs devoted to hired labor has steadily decreased over the last century, from more than 20 percent in 1910 to just over 10 percent in the agricultural census of 2002. The ethnic makeup reflects the greatest shift, from largely white or African American workers in 1910 to largely Hispanic workers today, an estimated 70 percent of whom are illegal immigrants.[4]

In the 2002 farm census the total number of farm operators was 3,045,318, with 2,128,982 principal operators. Until 1997, when responsibility for the farm census shifted to the Department of Agriculture, the Census Bureau had only one classification—farm operator—and this was limited to one per farm. This meant that, in some cases, the operator was a corporate person, not a human. Even if a husband and wife, or two or more equal partners, fully shared the management of a farm, only one could be listed as the operator. In a sense, this all but obscured what had been, in the past, the vital role of family members or partners. The present farm census corrects for this, but it still maintains the one farm–one operator system for most calculations of farm income and expenses. In addition to principal operators, it designates second and third operators (in other contexts, there may be up to five operators on some farms). The number of second and third operators in 2002 was 916,336, with the larger group designated as second operators. It is clear that most of these second operators were spouses, predominantly women (the number of female principal operators was only 237,819, but 822,093 were listed as second or third operators). This may have reflected farm wives' demand for recognition, but it also had a more self-serving purpose. When wives were listed as second operators, this in effect doubled the cap on the amount of federal payments to a given farm. The third operator category was smaller but in most cases included sons or daughters, who usually lived in separate households but worked on or helped manage the farm (the average age of third operators in 2002 was around twelve years younger than that of principal operators).[5]

It is impossible to determine the actual number of hired farmworkers, particularly as full-time equivalents. The census of 2002, based largely on reports by farm operators, lists just over 3 million hired workers, with 928,000 working as much as 150 days a year, or half time. This leaves 2,109,000 who worked less than 150 days, which might include someone who worked only one day. Also, these numbers undoubtedly include some double counting, particularly among illegal immigrants. Each quarter, the Department of Agriculture issues a report on farm labor, with its best possible estimate of the number at work in a given week. From April 8–14, 2007, American farmers directly employed 720,000 workers and, through contracts, indirectly hired 241,000, for a total workforce of 961,000. Since this was during the planting season, it probably comes close to or slightly exceeds the annual average. Of these workers hired directly by farmers, 580,000 worked more than 150 days a year, and 140,000 less than 150 days. The average wage for all hired workers was $10.17 an hour, a figure that reflects the salary (converted into hours and wages) of relatively few managerial employees (55,372 in 2002). For field work, it was $9.41; for livestock-related work (much of it on very large dairy farms), it was $9.55. This does not include the value of housing and meals, which are sometimes provided free, particularly for migratory workers (more than 45,000 farms, often among the largest and most productive, hired migratory workers in 2002). By law, all farmworkers are guaranteed minimum wages, and both federal and state regulations set reasonably high standards for housing. Still, farmworkers are among the poorest in America, with only domestics receiving comparable wages. Farm wages have traditionally been roughly half those in manufacturing, with more than half of all families living below the poverty line. Because of linguistic difficulties and the fear of deportation among illegal workers, farmworkers have not been in a strong bargaining position to demand higher wages and benefits.[6] Today, one of the reasons for higher farm incomes and low food prices is the meager wages paid to seasonal employees. The venerated hired hand of the past has been replaced by a group of nameless, non-English-speaking laborers. Labor has become almost as impersonal as machines.

If one adds the total number of farm operators and full- and part-time hired laborers, more than 6 million people work on or manage farms each year. Most of these are part-time workers, and most do not live permanently on farms, including a surprising number of operators (almost 500,000). Second operators, mostly wives, usually live in the same household as the principal operator. Thus, the most prevalent estimate of the total number of people now living on farms is around 4.5 million. This estimate has leaky boundaries on all sides. Many farm operators live on very small farms, have full-time off-farm employment, barely reach the $1,000 threshold in sales, and thus scarcely consider themselves to be farmers at all. The existing criteria exclude many people who provide vital services to farmers, such as the owners of large tractors or combines who make a living entirely by custom work on farms, or crop dusters who use airplanes to spread pesticides.

FARM INCOME

What about the income of farm operators? Once again, confusion reigns. A vast majority (more than 80 percent) of farm households have some-one working off the farm, sometimes accounting for the largest propor-tion of annual income. In many ways, the one development in the last century that has most affected the well-being of farm households has been the increasing availability of local jobs in manufacturing and service industries. Thus, the total household income of people whom the De-partment of Agriculture lists as farmers has steadily closed the earlier gap between farm incomes and other incomes in the economy. Sometime in the 1990s, farm incomes surpassed the national average. Also, despite the increased concentration of farm production and sales in fewer and fewer large farms, the gap between the household income of small and large farms has narrowed. Today, total farm income from all sources is more equal than in any other sector. The poverty rate among farm households is now below the national mean, even though it is highest among the more than 90 percent of rural Americans who are not farmers. Finally, the average wealth per household is higher in farming than in any other sector, largely because of the value of land and machinery.[7]

Obviously, off-farm employment explains this leveling of incomes. In some cases, small farmers chose to remain on farms and do some farm-ing, even though they held full-time jobs in a nearby city. In other cases, successful professionals and businesspeople chose to buy land, which is still a powerful mark of status in the United States, and to operate farms as a source of supplemental income, as a hobby, or for the tax break. Very high-income hobby farmers help raise the average income of farm-ers in the lowest 20 percent of production and sales. According to an estimate by the Department of Agriculture, the average 2007 household income of farmers was expected to be $81,588, but 85.9 percent of this would be derived from off-farm employment or other income sources. Only commercial farm households (those with farm sales over $250,000) would earn a majority of household income from farming operations, and for this category of farmers, total household income would approach $200,000. Thus, income levels among those who meet the very inclu-sive definition of a farmer tell us very little about the income derived from farming. I might add that tax returns tell us even less about farm-produced income, for tax breaks, tax avoidance, and a high level of underreporting of income lead to a dismal, and very misleading, portrait of farm income.[8]

Most data on farm-based income is assembled by the National Agricultural Statistics Service. The last full survey available as I write is the farm census of 2002 (the results of the 2007 census are not yet in). Almost all data and analysis involve cash income received by farm operators. The net cash income is simply the value of the total agricultural product and government payments less farming expenses. These data leave out capital gains on land, which the Statistics Service does not estimate, as well as several farm-related sources of additional income, which it does calculate. It refers to cash and farm-related incomes together as net farm income. Unfortunately, it is all but impossible to determine an accurate figure for the net income from farm-related sources, and I doubt that many farmers themselves could give a very clear accounting. This category includes income from custom work on other farms, rent for land leased to other farmers, timber harvested from the farm, and fees for hunting and fishing privileges. What once was a large category—the amount of products consumed at home—has shrunk to less than 1 percent, but another has remained important—the imputed rental value of the house lived in by the farm operator. This free rent may add 5 percent or more to the annual income of a small farmer. In 2002 the estimate of gross income from such farm-related sources was $5,689,226,000. To some extent, it is still true that the overall welfare of farmers is better than that indicated by what they produce and sell, a gain that is partly balanced by often high land taxes.[9]

Farmers have always suffered from price instability and thus large annual swings in income. Modern governmental policies have helped smooth these cycles, but not eliminate them. After the revolutionary farm bill of 1996, farm prices declined, but they have risen erratically since 2000. From 2005 through 2007, farm prices, gross income, and net income were higher than at any time since the mid-1970s. The value of the total product has risen rapidly since 2002, from around $225 billion to more than $300 billion in 2007, boosted by record-high corn prices (the ethanol effect) and near record-high soybean and wheat prices (caused in part by farmers' diverting acres from these crops to corn). Because of a developing worldwide shortage, wheat prices rose to more than $10 a bushel in late 2007, an all-time record in current dollars, and second only to the mid-1970s in real dollars. In March 2008 some wheat futures rose to $15 a bushel. Livestock prices are also high but are balanced in part by higher prices for feed grains. The Department of Agriculture estimated that the net farm cash income for 2007 would be $87.5 billion, higher than the 2006 income or the average for the preceding ten years. This is a boom time for grain farmers, but they know, from past experience, that it may not last.

Based on the 2002 farm census, farm income was then only slightly below the average of gross and net incomes over the preceding decade (table 2 offers a summary of these data). I use this survey to suggest certain aspects of farm income and the status of various categories of farmers (the 2002 figures should be more representative than the forthcoming 2007 census, which will come in the midst of a price boom). The Department of Agriculture classifies all operators of farms that sell products worth more than $250,000 a year as "commercial farmers." In 2002 these numbered 159,794. They received just over $160 billion of the total $207 billion in farm sales and government payments for that year, or more than 76 percent of the total (those with more than $1 million in sales received 46 percent). These commercial farmers make up an elite group in American agriculture. They are the most politically involved, exert the most influence on farm legislation, and are alone among farm groups in obtaining a majority of their household income from farming (in 2007 almost 70 percent, or approximately $140,000 a year). The operators in the next lowest category (farms with sales from $100,000 to $249,000), on average, receive less than a fourth of their income from farming. These 262,831 intermediate farms produced only 12.6 percent of the total product in 2002.

The concentration at the top is remarkable or, some would argue, scandalous. This is dramatically illustrated in figure 13 (page 163). Only 322,625 farms (those with sales over $100,000) sold 89 percent of the total product, leaving a measly 11 percent to the other 1.8 million farms. If farmers with annual product sales under $250,000 did not rely primarily on off-farm work, the income disparities in American agriculture would be extreme, and class resentment would be intense. Note that there are a good number of small farmers (annual sales under $100,000) who report farming as their only occupation, but if they did not have family members working off-farm, their household income would be below the poverty level.[10]

The overall net cash income of farmers in 2002 can be very misleading, for the 85 percent of low-production farmers are included in the overall averages. For farms that made a profit in 2002 (993,861, or less than half), the average net gain (or income) was $56,679, with more

Table 2. Market Value of Agricultural Products Sold Including Landlord's Share, Direct, and Organic: 2002 and 1997. (2002 Farm Census)

Item		2002	Percent of total in 2002	1997
Total sales	farms	2,128,982	100.0	2,215,876
	$1,000	200,646,355	100.0	201,379,812
Average per farm	dollars	94,245	(X)	90,880
By value of sales:				
Less than $1,000	farms	570,919	26.8	415,952
	$1,000	63,223	(Z)	76,813
$1,000 to $2,499	farms	255,639	12.0	277,074
	$1,000	422,136	0.2	461,290
$2,500 to $4,999	farms	213,326	10.0	265,667
	$1,000	762,554	0.4	951,581
$5,000 to $9,999	farms	223,168	10.5	267,575
	$1,000	1,577,184	0.8	1,897,337
$10,000 to $19,999	farms	197,967	9.3	228,297
	$1,000	2,781,507	1.4	3,218,769
$20,000 to $24,999	farms	58,190	2.7	65,342
	$1,000	1,285,921	0.6	1,450,207
$25,000 to $39,999	farms	109,310	5.1	123,854
	$1,000	3,438,976	1.7	3,909,475
$40,000 to $49,999	farms	48,596	2.3	55,775
	$1,000	2,154,772	1.1	2,481,435
$50,000 to $99,999	farms	140,479	6.6	163,510
	$1,000	10,024,295	5.0	11,716,046
$100,000 to $249,999	farms	159,052	7.5	190,919
	$1,000	25,401,608	12.7	30,406,176
$250,000 to $499,999	farms	81,694	3.8	91,503
	$1,000	28,530,105	14.2	31,848,311
$500,000 to $999,999	farms	41,969	2.0	43,973
	$1,000	28,944,401	14.4	30,138,101
$1,000,000 or more	farms	28,673	1.3	26,435
	$1,000	95,259,672	47.5	82,824,271
$1,000,000 to $2,499,999	farms	20,724	1.0	19,552
	$1,000	30,618,426	15.3	28,603,612
$2,500,000 to $4,999,999	farms	4,611	0.2	4,084
	$1,000	15,700,332	7.8	13,850,777
$5,000,000 or more	farms	3,338	0.2	2,799
	$1,000	48,940,914	24.4	40,369,882

(continued)

Table 2. *(continued)*

Item		2002	Percent of total in 2002	1997
Value of sales by commodity or commodity group:				
Crops, including nursery and greenhouse	farms $1,000	944,656 95,151,954	44.4 47.4	1,115,959 100,668,794
Grains, oilseeds, dry beans, and dry peas	farms $1,000	485,124 39,957,698	22.8 19.9	(NA) (NA)
Tobacco	farms $1,000	56,879 1,616,533	2.7 0.8	93,330 2,922,590
Cotton and cottonseed	farms $1,000	24,721 4,005,366	1.2 2.0	33,594 6,227,116
Vegetables, melons, potatoes, and sweet potatoes	farms $1,000	59,044 12,785,898	2.8 6.4	(NA) (NA)
Fruits, tree nuts, and berries	farms $1,000	107,707 13,770,603	5.1 6.9	93,784 12,883,386
Nursery, greenhouse, floriculture, and sod	farms $1,000	56,070 14,686,390	2.6 7.3	(NA) (NA)
Cut Christmas trees and short-rotation woody crops	farms $1,000	14,744 399,848	0.7 0.2	(NA) (NA)
Other crops and hay	farms $1,000	359,262 7,929,618	16.9 4.0	(NA) (NA)
Livestock, poultry, and their products	farms $1,000	1,094,608 105,494,401	51.4 52.6	1,322,054 100,711,018
Poultry and eggs	farms $1,000	83,381 23,972,333	3.9 11.9	75,444 23,389,462
Cattle and calves	farms $1,000	851,971 45,115,184	40.0 22.5	1,121,003 40,927,182
Milk and other dairy products from cows	farms $1,000	78,963 20,281,166	3.7 10.1	102,790 19,097,293
Hogs and pigs	farms $1,000	82,028 12,400,977	3.9 6.2	112,377 13,833,370
Sheep, goats, and their products	farms $1,000	96,249 541,745	4.5 0.3	(NA) (NA)
Horses, ponies, mules, burros, and donkeys	farms $1,000	128,045 1,328,733	6.0 0.7	(NA) (NA)
Aquaculture	farms $1,000	6,653 1,132,524	0.3 0.6	(NA) (NA)
Other animals and other animal products	farms $1,000	29,391 721,738	1.4 0.4	(NA) (NA)

(continued)

Table 2. *(continued)*

Item		2002	Percent of total in 2002	1997
Value of landlord's share of	farms	132,567	6.2	(NA)
total sales	$1,000	4,567,940	2.3	(NA)
Value of agricultural products	farms	116,733	5.5	110,639
sold directly to individuals	$1,000	812,204	0.4	591,820
for human consumption				
Average per farm	dollars	6,958	(X)	5,349
By value of sales:				
$1 to $499	farms	32,420	1.5	33,537
	$1,000	6,645	(Z)	6,949
$500 to $999	farms	19,145	0.9	19,013
	$1,000	13,124	(Z)	12,997
$1,000 to $4,999	farms	42,660	2.0	39,477
	$1,000	93,611	(Z)	86,410
$5,000 to $9,999	farms	9,598	0.5	8,293
	$1,000	64,517	(Z)	55,600
$10,000 to $24,999	farms	7,256	0.3	5,903
	$1,000	108,766	0.1	89,000
$25,000 to $49,999	farms	2,831	0.1	2,224
	$1,000	96,322	(Z)	75,614
$50,000 or more	farms	2,823	0.1	2,192
	$1,000	429,220	0.2	265,251
Value of certified organically	farms	11,998	0.6	(NA)
produced commodities	$1,000	392,813	0.2	(NA)
Average per farm	dollars	32,740	(X)	(NA)
By value of sales:				
$1 to $999	farms	3,297	0.2	(NA)
	$1,000	1,258	(Z)	(NA)
$1,000 to $9,999	farms	5,408	0.3	(NA)
	$1,000	18,562	(Z)	(NA)
$10,000 to $24,999	farms	1,344	0.1	(NA)
	$1,000	20,752	(Z)	(NA)
$25,000 to $49,999	farms	729	(Z)	(NA)
	$1,000	25,100	(Z)	(NA)
$50,000 or more	farms	1,220	0.1	(NA)
	$1,000	327,140	0.2	(NA)
(NA), not available; (X), not applicable; (Z), less than half the unit is shown.				

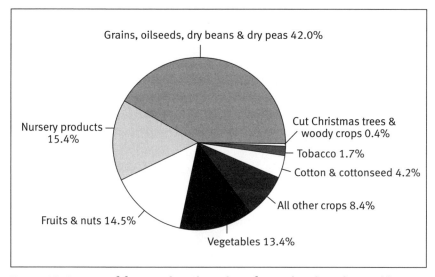

Figure 13. Percent of farms and market value of agricultural products sold: 2002. (2002 Farm Census)

than 82 percent of these gains on farms (212,965) having profits in excess of $50,000, which in most cases were commercial or high-level intermediate farms (more than $150,000 in sales). Balancing this were the 1,134,879 farms that suffered losses. Thus, for farmers as a whole, the total net gain was $40,514,055, or just over $19,000 per farm. Since some farms had more than one operator, as well as more than one household, the average net income of operators was $15,848. Clearly, only the ablest, largest, and most efficient farmers are able to make a living on the returns from their farms, even if one includes the unrealized capital gains from rising land values.[11]

This summary of income data provides a very positive view of the status of at least the key producers among American farmers. They are doing well, and as debates on a 2007 farm bill demonstrated, most are quite happy with existing farm policies and have fought against major changes (or what critics describe as major reforms). At the same time, most farmers with sales under $100,000 seem to be prospering, if household incomes are any indicator. They are better off than most Americans. This is because of off-farm jobs, along with either small profits or at least tax benefits from farming, and increased wealth because of rising land values. But even from the perspective of farmers, all is not well. And from the perspective of many outside critics, contemporary American agriculture is a scandal.

Critics and Criticisms

Everyone has to concede one point: American farmers have achieved a level of efficient food production unprecedented in world history. It seems almost unbelievable that 322,000 principal farm operators, on 322,000 farms, guide the production of 89 percent of all domestic foods and fibers (cotton, wool) consumed in the United States, with a remarkably small supply of family or hired labor. They also produce enough food and fiber exports (in excess of imports) to supply the other 11 percent, at least in quantity if not in balance of products. The principal operators of these 322,000 farms make up only slightly more than 0.001 percent of the population of the United States.

Economists point out that different agricultural policies might have slightly increased this level of efficient production. Even most large farms have not achieved the maximum possible economies of scale. The race toward more specialization, larger farms, and fewer farmers is not over. On the other side, it is clear that different policies would have lessened the efficiency of production. This includes major subsidies to keep small farmers in business. Less efficient production would mean a larger share of the American labor force employed in agriculture and a proportionately lower amount in manufacturing and services, along with a lower average level of consumption for everyone. Also, lower food production in the United States could have had a dire impact on world food supplies and on efforts to eliminate world hunger.

One can cite many possible benefits from policies that would have enabled more Americans to remain on farms. What one believes "beneficial" will always reflect personal taste and values. More small farmers would have avoided dislocation and, at times, migration. This would have meant, over the long term, a qualitative gain for many farmers and possibly for the nation as a whole because of the vast social problems created by displaced farmers, particularly those from the South. After all, the greatest surplus problem in American agricultural history has involved not commodities but people. It is conceivable that a nation of smaller family farms would have benefited the environment, but this begs specifics. Certainly, smaller farms would have required more land in cultivation and thus more forests cut or not allowed to regrow, but possibly fewer pesticides and chemical fertilizers, less distant transport of foods, and less burning of fossil fuels, although one can dispute each of these claims. One could argue that the quality of foods would have remained higher with smaller farms and the growing of more flavorful if not more shippable varieties. Also, more and smaller farms would have helped in the survival of small rural towns and villages in remote areas of the country.

Most critics of present-day American agriculture focus on federal policies that helped produce the present size and efficiency of a few farms and policies that now subsidize even the largest and wealthiest farmers. Governmental policies helped create the oligarchy at the top and now help perpetuate it. Since Congress seems to spend most of its time catering to various powerful economic interest groups, it has given large commercial farmers (the Department of Agriculture definition) almost everything they want. This worries critics on two counts: economic fairness, and the effects of such a subsidized agriculture on the environment, food safety, animal welfare, nutrition, the fate of small farmers, and the health of rural communities and small towns. Mixed in is a sense of loss as the traditional farm is passing away, and with it the noncommercial, nonconsumer values that Americans purportedly once held dear. The "soul" of America is thus at stake.

Much of the critique of commodity policies come from experts within the agricultural establishment. Most notable is Willard Cochrane, who has written detailed books on the various farm programs, worked on farm bills during the Kennedy administration, and has proposed various reforms. He, like so many critics, deplores the amount of subsidies for the larger crops and for the large and affluent farmers who need them least. He has suggested more support for small farmers and for healthy fruits, nuts, and vegetables (since more than half these products come from irrigated farms, it is a mistake to argue that they have not received any federal support). He has also proposed more federal funds for conservation, environmental protection, food safety, and the rural poor. Consistent with proposals that go back to the Brannan Plan in the Truman administration, he wants to base support payments not on past production records but on the need for income stability among individual farmers. In at least small ways, all these proposals helped shape the debates on the proposed farm bill of 2007.[12]

But such reforms, if one so conceives them, will be difficult to achieve, given the present political clout of commercial farmers. The Department of Agriculture, a now engorged bureaucracy, has many functions, some of which are scarcely tied to farming (the Forest Service, rural development). But at its core, it is the one agency that represents successful farmers. Its programs helped create these farmers and they remain its main constituency. In the same way that the Department of Commerce represents the interests of business enterprises, the Department of Agriculture represents the most productive farmers. Beginning with the Resettlement Administration and the Farm Security Administration in the New Deal, the Department of Agriculture has exerted some effort to aid small farmers (and, only recently, minority farmers) through rehabilitation loans, tenant-purchase plans, and enlarged payments. The proposed farm bill of 2007 was full of gestures to these constituencies, with almost every program offering special subsidies to "beginning and socially disadvantaged" farmers. But there is a bit of irony here, for if these presently disadvantaged farmers make good, or even survive as full-time farmers, they will do so by gaining the demanding managerial skills needed in contemporary agriculture and, in most cases, by developing large-scale, highly capitalized farms. They will thus become part of the ruling elite. A few may make a good living in niche farms, or what I call "alternative agriculture." But most are doomed to failure, and continued subsidies to them will be a form of welfare, not an aspect of industrial policy.

Race is now a prominent issue in farm politics. This is a bit ironic, because the concern for African Americans, in particular, comes much too late. Most blacks have long since departed agriculture. Many reasons lay behind their exodus, but one was the unfair treatment they received from the federal government, beginning with the Morrill Act of 1862. In the South, they were excluded from the large, segregated land-grant colleges and had to make do with small, underfunded, academically inferior agricultural and mechanical schools for blacks. The outreach programs from the land-grant institutions also discriminated against blacks, with only a few black extension agents to serve their needs. Beginning in the New Deal, despite a long and bitter controversy in the Department of Agriculture, both sharecroppers in general and black sharecroppers in particular rarely received their legal share of payments to farmers. African American farm owners were rarely represented at all in the local committee system that determined allotments. And so it continued even after the Civil Rights Act of 1964, when most of the aggrieved blacks had already moved away from their former farms. The few remaining black farmers have successfully sued in the court and, in one case, won compensation for past injustices.[13]

Even without overt, or unintended, discrimination, most black farm owners had a limited chance of success in the competitive revolution after 1950. Most owned small farms, lacked the education of whites, had limited political clout, and could not compete with the large-scale, highly mechanized cotton, tobacco, and soybean farms or dairy and livestock operations that soon dominated southern agriculture. A head start, including the original base acreage won in 1933, usually determined later outcomes. In fact, in retrospect, it is amazing how federal policy from 1933 on ratified existing inequalities. Thus blacks, along with most tenants and farmers who owned small or poorly located farms, were almost doomed to failure unless they were able to gain off-farm employment. As it turned out, a much higher proportion of small white farmers continued part-time farm operations, combined with factory or service employment, than did blacks. This was a matter of choice, job discrimination, and the fact that such a small portion of blacks owned their farms.

Other issues cited by outside critics involve animal welfare and food safety and quality. Animal rights advocates are having some impact on consumer demand, if not on government policies. Of course, from a vegetarian perspective, the whole livestock enterprise is a story of enormous cruelty as well as the source of an unhealthy diet. But plenty of nonvegetarians protest the way we grow and kill animals. I need not dwell on the caged chickens that cannot roost normally and cannot even spread their wings, or the hogs kept in such small cages that they cannot turn around, or the calves that are taken from their mothers at birth and grown to veal size in cages and on artificial feed. One could also cite the huge, crowded feeding lots for cattle and the enormous confinement dairy farms where cows do not have access to pasture. I noted earlier the antibiotics routinely fed to chickens, the steroids given to almost all beef cows, and the hormones fed to dairy cows. These realities have led to a growing market, among those who can afford it, for free-range broilers and hogs, eggs from noncaged chickens, grass-fed beef, and organic beef and pork. Yet one has to acknowledge the obvious: these contested methods have tremendously lowered the price of meat proteins. So far, only a small minority of consumers is concerned enough about animal welfare to pay up to double the present price for less efficiently produced meats.

Safety and flavor are valid concerns about any food. Whatever the safeguards in the meatpacking process or in its transportation and sale, meat will always pose potential dangers to consumers. This is not a new problem. In fact, the meat available in supermarkets today is less likely to be contaminated with botulism or other deadly bacteria than that available in the past from local farmers or village meat markets. But many more people are at risk. Farmers boast (with justification, I think) that the food consumed by Americans is safer than at any time in the past. Never before in history have consumers had such a wide array of food choices, with almost every imaginable food, both domestic and exotic, available at any time of the year. Given the scope of the food distribution system, it is a marvel that more outbreaks of bacterial infection, like the one affecting spinach in 2006, do not occur. More rigorous inspection and tougher regulations may allow marginal improvements, but overall, our food is safe.

But is our food tasty? As an avid gardener, I deplore most store-bought tomatoes. They are hard and almost totally lacking in flavor. My favorite varieties of melons are too thin skinned for shipping at any distance. It is almost impossible to get fresh ears of corn that have the full flavor of those just picked from one's garden. Pineapples and other exotic fruits, except for bananas, suffer from being picked half green. Cold-storage apples rot too quickly when purchased and sometimes lack flavor. In local markets in Tennessee, I often have to buy inferior California peaches because they are the only ones available; the more flavorful southern varieties are nowhere to be found. And so it goes. The trade-off for having access to almost any fruit or vegetable at any time of the year and at an affordable price is inferior flavor. One can shop in specialty stores and, for a price, buy very fresh foods shipped overnight by air, but this is largely a luxury enjoyed by expensive restaurants. Buying locally grown fruits and vegetables in season at farmers' markets is one way to get better-tasting produce, but most Americans are addicted to year-round availability. My assessment is more positive for frozen foods. In most cases, flash-freezing by commercial processors preserves flavor and texture better than home freezing.

AGRICULTURE AND THE ENVIRONMENT

What about farming and the environment? This involves three issues: the inevitable effects of any type of farming on the environment, the added environmental effects of traditional American agriculture, and the new environmental hazards created by modern agriculture with its large machines and abundant chemicals.

In the jargon of economists, almost any economic enterprise involves unpaid externalities, most of which are environmental in nature. In the case of agriculture, farmers in the past rarely had to pay for the necessary or unnecessary damage they did to the soil, air, and water, damage that inflicted a burden on the larger public. Of all human activities, the cultivation of crops has had the largest impact on the face of the earth, beginning with the elimination of up to half of all forests. If one places a high value on an environment little affected by humans, then agriculture by necessity is hostile to environmental health. It has eliminated wilderness, shifted the balance of plant and animal species, altered the hydrological cycle, and, in a limited way, altered climate. But whether these effects are all good or bad depends on one's values. For example, the orderly, green, garden-like fields of England or the Netherlands may appear, to some at least, to be an improvement on wild nature. In any case, these built-in effects of farming were necessary for the growth of human populations and the types of civilization that ensued from that increase. Today, any strategy to eliminate crop farming and domesticated livestock would have to include the means of eliminating most humans now on earth, for only a small remnant could live by hunting and gathering.

Thus, the only environmental costs that are pertinent to present-day policy choices are those that are open to mitigation by farmers without impossibly high economic costs or, in some cases, such as soil erosion, with major economic gains. Soil is the one resource most closely tied to farming. Erosion and nutrient depletion are the greatest threats to soil. In America, from colonial times to the Great Depression, American farmers as a whole were reckless in their use of the land. So much land was available that they could afford to deplete soil nutrients or let hillsides erode with little consequence, for they could clear new land or settle in new areas. Often, in the short term, they gained extra income by making such extravagant demands on the land, even as they benefited by a rapid, often wasteful, clearing of trees. But they incurred high future costs for the larger society, costs that began to come due, in a major way, by 1930.

It may surprise many people (or disturb those who are most critical of contemporary agriculture) to learn that almost all evidence suggests that American soils are now eroding at a slower rate than at any time in the last two centuries. This was not true as recently as the mid-1980s. By then, the conservation policies enacted in the New Deal had eliminated the worst forms of erosion. The deep gullies of the Southeast, the blowing dust of the Great Plains, did not reappear. But the boom of the 1970s, the push of grain farming onto new and more erodible lands, and the relative neglect of conservation issues by too many farmers created a situation in which the actual soil loss each year rivaled that of the drought-stricken 1930s. This loss led to new and more vigorous efforts to control erosion and a gradual improvement over the next two decades.

Erosion is part of nature. In time, all soil, and even all rock, will erode. Such loss is compensated for by tectonic forces that create new rock and, eventually, new soil. But clearly the cultivation of crops, or the overgrazing of grasslands, hastens erosion by wind and water. Because of recent conservation efforts, well over half of American cropland is now free of the degree of erosion that threatens sustainable production—that is, the amount of soil lost per year per acre is less than the two to six tons (the exact amount varies with soil types and vegetation) that could seriously threaten production over the next century. But there are major exceptions to this good news, for around 10 percent of soils are still quite vulnerable to erosion. At the same time, the amount of land in cultivation was stable or even declining until the recent ethanol boom, and the amount of forestland has increased over the last hundred years. The old American custom of clearing new ground is all but over, even as the enormous annual demand for wood as fuel is history.[14]

The natural supply of nutrients, and the level of organic matter, is at an all-time low in most soils that are intensively cultivated. Only a few traditional, or organic, farmers still carry out the old nutrient replenishment tactics, such as a complex pattern of crop rotation, the use of legumes for soil building, or the use of manure. Commercial farmers buy fertilizer instead, which so far costs less than the loss of production required by past methods. From the public's perspective, this present practice poses two risks: future scarcity of fertilizers, and the pollution that results from the overuse or careless use of synthetic fertilizers. If fertilizers became too expensive for profitable use (since nitrates are so dependent on methane, this is not an impossible scenario), farmers might have to return to older means of preserving soil fertility, including the planting of cover crops. This would reduce the amount of land devoted to grains, beans, and cotton; raise the price of such basic crops; and risk more food scarcities in underdeveloped areas of the world. In effect, synthetic fertilizers are now necessary to feed the world's population of more than 6.5 billion people.

Because of the heavy use of synthetic fertilizers and pesticides, water pollution has become an even greater environmental risk than erosion. In any form of cultivation, some soil content ends up in streams and lakes, either by erosion or by the leaching of phosphates and nitrates into the subsoil and eventually into groundwater. This is less of a problem with no-till or low-till farming. Thus, the majority of nonpoint water pollution in most of the world derives from agriculture, followed by household wastewater. Nonpoint sources are small but widely distributed sources of pollutants that are difficult to identify and control, unlike the point-specific pollution of a factory or a large hog or poultry farm. The normal flow of soil nutrients into groundwater is increased by the use of fertilizer, and particularly its overuse. This is an ever-present risk with nitrogen, which leaches quickly from the soil. So if farmers apply nitrogen-containing fertilizer before the time of rapid plant growth, they almost ensure the loss of much of the nitrogen before the crop absorbs it. Too much nitrogen in groundwater can reach toxic proportions. A heavy load of both phosphates and nitrates supports the rapid growth of algae in streams and lakes. This in turn strips oxygen from the water, which can lead to fish kill and hastens the filling in of marshes and speeds up the gradual death of lakes (eutrophication). The only defense against these effects is a better-calibrated application of fertilizer, and even this cannot fully solve the problem because of such variables as the amount of rainfall. For this reason, modern fertilizer-based agriculture has a greater pollution risk compared with traditional farming. The rotation of row crops with grasses allows a recharging of soil nitrates through decaying vegetation, and the planting of nitrogen-fixing legumes adds nitrogen from root nodules. This organic nitrogen is only slowly converted to the inorganic, soluble nitrates that plants can absorb and thus poses much less of a threat to groundwater than synthetic fertilizer.

In the next two decades, water scarcity may be the greatest threat to American agriculture. Irrigation accounts for more than 80 percent of all fresh water utilized in the United States. More than half the crops west of the 100th parallel use irrigation water drawn from dammed rivers or from aquifers. By some estimates, at least 25 percent of water used for irrigation is not replenished annually. This is most true for the Ogallala aquifer in the Great Plains, which is shrinking each year and may, in the next three decades, be all but exhausted. In the Colorado River watershed, urban demands are already affecting agricultural users, some of whom have sold their water rights to nearby cities. By 2007, several years of drought had shrunk by half the two largest Colorado reservoirs—Mead and Powell—and it is feared that they may never fill again. Global warming is already shrinking mountain glaciers and the amount of snow cover that remains into the summer to feed streams. In the past, western farmers enjoyed cheap, federally subsidized water. In the future, they may have to use water more efficiently (covered canals, drip systems), which will add greatly to costs.

Earlier, I enumerated the possible problems posed by the use of antibiotics and hormones in livestock. Insecticides, fungicides, and herbicides have the potential to harm those who apply them or possibly a larger population if residues remain on foods or spread beyond the fields. However, few certified pesticides leave toxic wastes in the soil. Agriculture can rarely be blamed for toxic waste sites, with one exception—large lagoons for the storage of poultry, hog, or cattle manure, which some-times leak or burst in floods and spread toxic materials into streams, kill-ing fish or endangering public water supplies. North Carolina has had the worst problems with such manure disposal.

Many of the environmental problems created by farming are shared with the larger society. Agricultural machines, powered by fossil fuels, contribute to air pollution in the same way automobiles do, and to the greenhouse gases that boost global warming. In the area of air pollution, however, farmers have a unique problem—the methane expelled by ru-minants (particularly our large population of cows) and, to a lesser ex-tent, the methane generated by manure disposal and flooded rice fields. Together, these contribute nearly a third of methane from human origins, or nearly 3 percent of the greenhouse gases emitted in the United States. Unless Americans become vegetarians, there is no way to eliminate this methane production, although it may be possible to collect the methane from manure lagoons and use it for energy.

Despite the problems caused by large cow feeding lots and hog and poultry factories, they are at least easily identified as sources of pollu-tion. This is not true for another worrisome problem endemic to grazing livestock. There is no way to prevent their excrement from percolating into groundwater. Also, except in a few counties, a large proportion of these animals have direct access to small streams and creeks. That is where they get their drinking water and where, in the hot summer months, they congregate to cool off. The result is trillions of E. coli bacteria in our streams. A few strains of E. coli are toxic to humans, causing a range of symptoms from diarrhea to intestinal bleeding and even death. These bacteria can directly affect the meat supply, as in the hamburger scare of 1993, or get into the water supply used to irrigate or wash vegetables, as in the spinach scare of 2006. Present efforts to pass local laws requiring the fencing of all creeks, at great cost to farmers, would solve only part of this problem. The burden falls largely on the processors of meats and vegetables. When E. coli infects springs and wells—the sources of drinking water for many rural people—there is no easy solution. Beyond the hu-man cost is the problem of dead streams that support no fish.

The livestock problem is more widespread than any other environ-mental issue tied to farming. This is because 108 million cows, sheep, horses, and goats reside on over 1 million farms, or almost half of all farms. Of these, more than 95 million are cows. Unlike so many other aspects of farming, cows are still widely distributed. It seems that almost all farmers, including small and hobby farmers, have some beef cows at pasture. In much of the West, the land is suitable only for grazing. In much of the country, range cattle remain outside all year long, and their manure helps recycle nitrogen. This is a blessing, except for groundwater seepage. In the past, cows were usually housed and fed in barns in the winter, with straw bedding to soak up both urine and manure. When scattered on fields in the spring it was a wonderful source of nutrients. This traditional method is almost gone, and manure, once a great asset on farms, is now more of a problem.[15]

This concludes my summary of the present status of farming in the United States and of the various challenges faced by farmers. These issues are continually being addressed by Congress and by lobbyists from many different interest groups, and they are often of paramount concern among active farmers. They will remain at the center of public policy debates, as they were in the discussions of the proposed 2007 farm bill. But one assumption of most of these debates is that our overall system of farming, which developed over the last century, is beyond challenge. Thus, the focus has been on internal adjustments to and reforms of the existing agricultural establishment. This suggests a final challenge to my effort to write an update on American agriculture. Can one imagine a different agricultural system for the United States? Are there viable alternatives to the way we grow our food? This is the subject of the next chapter.

8. Alternatives

Are there workable alternatives to the dominant agricultural system in the United States? Thousands of people believe there are, and they are hard at work promoting what they see as necessary substitutes for the present large-scale, specialized, mechanized, chemical-intensive farming that produces most of our food and fiber. These critics of the present system may agree on what is wrong with the present system, but they advocate different prescriptions for its reform. The labels they use for these alternatives also vary—permanent, sustainable, regenerative, bio-dynamic, holistic, natural, ecological, organic, low-input. In this chapter, I survey some of these alternatives.

LONELY FARMERS

At least in a global perspective, some deeply ingrained aspects of farm-ing in America were soon ubiquitous, particularly those involving farm placement and ownership. The first English settlers in America came largely from a village-based agriculture. Homes and churches were lo-cated in a rather compact area, with fields and forests surrounding the village. This pattern prevailed briefly with the first settlers in Virginia and for at least a few decades in parts of New England. In New England some of the new towns even kept the open field system of East Anglia, with heads of families assigned areas of the commons to cultivate each year. The early New England towns, with more land than they could clear and cultivate, retained ownership of the forests and meadows, as well as a village commons (some of these still survive). But as the population in-creased, and maturing sons needed land of their own, towns made grants of land located farther away. Many of these new landowners built homes on their land, creating a mixed pattern of village dwellers and dispersed farmsteads. In time, the majority of farmers would live outside the town centers, with some too far away to attend church and school or even participate and vote in town meetings. Communal norms became more difficult to enforce, and the sense of community weakened.

In the southern colonies, a somewhat different pattern developed. There, large royal grants led to sales to individual farmers, who usually chose to live on their land. Widely scattered homesteads became the rule. Towns and villages were few in number, and any type of close communal life was almost impossible. This was also the pattern in most of New York and Pennsylvania. But in the Hudson Valley and in the South, a few very large landowners were able to acquire dependent workers (indentured servants or tenants in New York, servile workers in Maryland and Virginia), which led to clusters of dependents around the principal plantation home. Thus, by the eighteenth century, the typical American farm family lived in near isolation in its own small kingdom, with very little external control over its operations.

The geological survey after 1794, and the township pattern it created from Ohio westward, reinforced this pattern of settlement and an agricultural system of widely dispersed farmsteads. In the nineteenth century, as settlements moved into the more arid Great Plains, wheat farming or, farther west, the grazing of cattle led to farm and ranch homes that were a mile or more apart. European visitors were often amazed at what they perceived as the lonely lives of American farmers. Some eastern reformers lamented the antisocial aspects of such settlement patterns, and European immigrants, particularly women, suffered from their lonely lives on the plains. Only in the twentieth century did improved roads and automobiles lessen the isolation. Ironically, today the patterns are shifting again. Unregulated suburban sprawl and strip development along highways have blended urban and rural settlement patterns, even as an integrated employment market has allowed most rural Americans to work for urban or suburban firms.

The most influential critic of dispersed farms was economist Henry C. Carey. He was a friend of Ralph Waldo Emerson, an adviser to Abraham Lincoln, and one of the founders of the Republican Party. He eventually rejected free-market theory and free trade. The son of an Irish immigrant, Carey had an animus against the United Kingdom and its new factory system that had reduced low-paid workers to something close to slaves. He wanted tariffs high enough to foster American manufacturing and stressed consumer demands and mild inflation as engines of economic growth.

Carey absorbed some of the ideas of Comte Claude Saint-Simon and sought a richer communal or associative life for Americans. From his perspective, farming in America was a major scandal. As he put it, we had no agriculture, for our system lacked anything close to culture, in the sense of either artful engagement or the maintenance and improvement of land. Instead, we had a system of land mining that gradually degraded the great soil bank on which life depends. Farmers used up the soil in the East and then moved west to lonely, isolated farmsteads. He pitied them. There, they too often grew commodities for export, such as wheat and cotton, or they grew corn to produce pork and beef (Carey, a near vegetarian, believed such animal fats were harmful to health). In either case, they drew down on soil nutrients, with only limited recycling of waste products, which he collectively referred to as manures. He particularly resented shipments of grain abroad, for these represented nonrecoverable withdrawals from our soil. Equally wasteful were large commercial cities, which consumed the products of the soil without any recycling at all.

Carey believed that a scientific agriculture, with better tools, would enable farmers to prosper on smaller and smaller plots of land. They could live in or close to villages, enjoy good schools, and find vocations that fit their varied skills and talents. He applauded the type of localism that reformers today call bioregionalism, with small factories interwoven with small farms and almost no external trade. He believed that it was possible, with the recycling of all waste products and the use of legumes, not only to preserve but also to replace the natural fertility of soils. His ideal economy was that of the Netherlands. With the steady improvement of our soil bank, and with farms dedicated primarily to vegetables and fruits, we could support an almost unlimited growth in population. And since people were the greatest economic asset of all, the more the better.[1]

ALTERNATIVES IN LAND TENURE

The other prominent feature of American agriculture, and one so appealing to European immigrants, was fee simple ownership of land, which Carey applauded. Except for lingering commons in New England, American governments quickly alienated land in behalf of rapid development. Of course, in a strictly legal sense, the sovereign entity still owned all land. The federal government and the states granted conditional land titles to individuals (by eminent domain, American governments could reclaim the use, but under our constitutions, only with due compensation to title owners). We thus eliminated the last remnants of feudal hierarchies, soon outlawing even entail or primogeniture laws. People who, in Europe, had lived as tenants or with long-term leases or had paid quitrents were now free of all such dependency.

This almost unrestricted control over land, and often even the minerals beneath the land, created an agricultural system that still exists today. But today the sovereign people, acting through their governments, have placed more and more restrictions and qualifications on the managerial prerogatives of title holders. Some reformers wanted even more regulation or planning (a socialization of management), while others repudiated private ownership in favor of communal or corporate property (a socialization of property). The push for regulation, often in the furtherance of environmental goals, continues to grow. Attempts to socialize land have largely failed, although a few isolated examples of communal agriculture have at least illustrated one alternative to the American agricultural system. I do not consider forms of institutional ownership (prison farms or county poor farms, monastic farms, university-owned demonstration farms) as alternatives.

I have written extensively on communal colonies. In the mid-nineteenth century the ideas of Robert Owen and Saint-Simon led to at least fifty socialist colonies in the United States. In the late nineteenth century the cooperative commonwealth movement, informed by Marxist ideas, led to a dozen or so rural colonies, only one of which survived for as long as thirty years. Some estimate that radical or countercultural youth formed more than 3,000 communes—most of them short-lived—in the 1960s and 1970s (a struggling few survive today). Yet none of these efforts came close to demonstrating a realistic alternative approach to farming. None lasted long enough to gain generational continuity, let alone any appreciable growth in membership. But there are three significant exceptions that do meet these criteria: the Shakers, who were doomed to near extinction after a century because of their practice of celibacy; the Amana Society in Iowa, which remained communal from 1855 to 1931; and the Hutterites. Since only the Hutterites survive today, I will use them as an example of communal success, just as I would use the kibbutzim of Israel, and not the coercive state farms of the former Soviet bloc, if I were surveying global agriculture.

The Hutterites were an Anabaptist sect that formed in 1536 in what is now Austria. Doctrinally similar to the Mennonites and the later Amish, they distinguished themselves by adopting a community of goods.

Though much persecuted, the Hutterites experienced a golden age in Moravia in the late sixteenth century. This ended in 1593, and the movement barely survived during the next 200 years. The Hutterites moved first to Slovakia, then to Transylvania, where only one colony survived by 1762. By 1771, a small remnant had settled among Mennonites in the Ukraine. There they were unable to continue their communal life and almost blended into the Mennonite population. In 1857 the Russian government permitted a small group of Hutterites to buy their own land and reestablish a bruderhof. Two more followed, and it was these three bruderhofs that established colonies in South Dakota beginning in 1874. They fled conscription laws in Russia and gained a firm promise from President Grant that they would face no conscription in the United States because there would be no wars (a promise broken in 1917).

In America, the three original Hutterite colonies, each of which retained some distinct practices, soon began to found daughter colonies. With strict prohibitions on birth control, the population doubled every fifteen years. By 1950, the Hutterites had the highest fertility rate of any known population (averaging more than ten children for each mother). After suffering hostility and oppression in World War I, most colonies moved to Canada, but some moved back after the war. They spread through the Dakotas and Montana and to three Canadian provinces. Today, they number more than 40,000, living in about 475 colonies (the number grows each year). By far, they are the most successful communal sect in either Canada or the United States.

The Hutterites try to follow the original Church at Jerusalem. They hold all property in common, live in colony-owned apartments, and, through most of their history and on most colonies, share common meals. They are complete pacifists and do not participate in politics. Unlike the Amish, they never rejected modern farming machines or methods, but they traditionally rejected all worldly attractions (no radio, television, or passenger cars, and only periodicals that relate to their farming business), restrictions that have recently broken down in some of the more liberal colonies. At first, on the sparsely populated Great Plains, they maintained their own public elementary schools, with outside but understanding teachers. Today, many have to attend non-Hutterite high schools, and a few attend Mennonite colleges. At first they were primarily grain and cattle farmers. But in time, land prices and hostility in Alberta made it difficult to gain enough land for daughter colonies. Thus, they turned to more intensive agriculture, such as poultry, and to small manufacturing. They continue to stress self-sufficiency, but they cooperate with extension agents and, in some cases, have moved ahead of small farmers in their use of technology. They face some resentment because their bulk purchases typically bypass local retail merchants. All is not idyllic. They have faced internal schisms, have struggled to resist assimilation near cities such as Winnipeg, and have a low per capita income. But the Hutterites continue to grow and have proven that, given a binding ideology, communal agri-culture can compete with individualistic, dispersed farmers. In the 1960s they became an inspiration to many countercultural rebels.[2]

AGRARIAN REFORM

Despite the limited appeal of religiously inspired communes or social-ist experiments with common ownership, one reform movement—agrarianism and closely related single-tax schemes—offered a widely ap-pealing challenge to the American system of landownership. By agrarian-ism, I mean schemes to provide broad access to land and legal methods to prevent land monopolies. For some reason, in the United States even historians have turned the word *agrarian* into a near synonym for *agricultural* and, in the process, obscured its radical implications. In much of the world the label still identifies the land reform policies pushed by tenants or peas-ants who want to gain access to land—to their share of a natural resource not created by human labor yet monopolized by large landowners.

Anglo-American agrarian reform efforts began in 1775 in England. Thomas Spence published a book on the real rights of man and soon formed the Society of Spencean Philanthropists. He correctly noted that a natural right to property, as clarified by John Locke, is the right of every individual. Every person is entitled to have access to and a share in the products of nature, of which land is the most important component. To provide such equality of access, Spence proposed that parish corpora-tions assume ultimate ownership of all land. They could not sell it, for land should not be a marketable commodity. Instead, they should rent the land to farm families and use the rent to fund all governmental services (a single-tax system). When these tenants died, the land would be avail-able for new farm families.

Thomas Paine accepted the basic ethical principle enunciated by Spence—that every person has a right to his or her share of the inherent or unimproved value of soil. In 1795 he wrote an essay titled "Agrarian Justice" in which he advocated an inheritance tax on all estates and the use of this tax revenue to create what amounted to a birthright fund. This fund would pay each maturing adult, at age twenty-one, the sum of money equivalent to the value of his or her rightful share of the nation's land. Thomas Jefferson bought into this principle and proposed that each child born in Virginia receive a fifty-acre birthright.

These agrarian proposals turned more radical in the 1820s, when organized urban artisans formed the Working Men's Party and gradually embraced agrarianism as a possible solution to their declining incomes. They faced competition from imported factory goods and from a surplus of workers in many trades. As proprietors, they wanted to escape the ser-vility of employment, or wage labor. In Philadelphia in 1828 they asked the federal government to reserve from entry and sale all the remaining public lands and to lease twenty to forty acres of public land free to each working family, or what they called their birthright. If these families oc-cupied and cultivated the land, they could enjoy it without cost for their lifetime. These men did not believe that anyone should "own" land or alienate it for loans, but they did believe that tenants rightfully owned any improvements they made to the land (which they had produced, whereas God had made the land) and could sell those improvements to new homesteaders. In 1829 Thomas Skidmore took these ideas to their logical conclusion. In one of the most radical books of the century, The Rights of Man to Property, he asked the state of New York, which he be-lieved was already cursed by monopolistic estates, to use a constitutional amendment to vacate all land titles and cancel all debts. Each person older than twenty-one would receive a credit instrument that vested in him or her the value of that person's share of all the land in New York. The state would use an auction to offer life-tenure leases to farmers, and anyone could use his or her certificate to bid on land. On the death of the origi-nal purchaser, the land reverted to the state. Only such a drastic policy, he believed, would uphold the right of every person to property.

Skidmore gained few disciples, but the Working Men's Party had un-leashed a powerful idea—free homesteads. One of its New York members, publisher George Henry Evans, launched the most influential agrarian movement in American history in 1844. He first called it the Agrarian League but soon changed the name to the National Reform Association (NRA). It gained a large following, with local chapters in every state. It had one major platform—that the federal government provide a free, exempt, and unalienable 160-acre homestead to every American family that wanted to go onto the land and cultivate it. These homesteads could not be sold or used as debt instruments. A second, less significant reform involved state laws to limit the amount of land owned by any individual (a bill to limit the amount to 300 acres almost passed in Wisconsin). The

NRA and sympathetic congressmen kept homestead bills before Congress throughout the 1850s and gained a partial victory in 1862 with the Homestead Act. Unfortunately (at least from the NRA's perspective), it did not make such homesteads unalienable. Homesteaders could sell their land, collect capital gains, and even buy up extra land, which in time would restrict ownership. The NRA faded rapidly after the Civil War, but the principle involved would gain its most influential advocate in the rightful successor and inverted namesake of George Henry Evans: Henry George.[3]

Henry George is rightfully considered the most persuasive advocate of a single tax in his famous book *Progress and Poverty*. He accepted the moral arguments of Spence and Evans and tried to turn the principle into a compelling economic theory. In effect, George tried to socialize access to land by means of the single tax. He believed that people deserved the fruits of their labor and thus full and free ownership of both capital and consumer goods. He loved a free, competitive form of capitalism, and inimical to this, he believed, was land monopoly. Because of inherent fertility or favorable location, land values rose with the growth of population and industry. But those who profited from such rising values did not deserve these unearned benefits, which gradually allowed landowners (in cities even more than in the countryside) to gain great wealth, even as an increasing share of nonlandowning workers fell into poverty. Since the unearned increments were relatively minor in farming areas well away from cities, the single tax would be small, usually much smaller than existing taxes on land and improvements. Thus, George believed that his single tax would benefit most farmers, particularly those who wished to enter farming. It would force wealthy owners to assume the major burden of paying for all governmental services. Such a tax meant that, in all areas, access to land would be, in effect, free, although in many cases property values would largely reflect the earned value of improvements. Capital and labor would be able to collect a full return for what they produced, in what would be a rent-free society.

The single-tax philosophy had broad influence, particularly in such European countries as Denmark. It seemed beguiling to many who read George's books. But it had a limited impact on land policies in the United States, where the large amount of underused land meant low agricultural rents, at least until the Great Depression. In several ways, his arguments are more applicable today, when land prices are high. In part they are high because of the nonagricultural demand for land, not only for suburban development but also for rural amenities, such as hunting, or even for tax breaks among the affluent. In metropolitan areas, farmland is often so valuable that it is not rational to continue farming, a problem that would disappear under a single-tax system. The opportunity costs are now so high that many farmers could sell their land, invest the returns, and earn a higher income than that obtained from farming. But even in more remote areas, land prices are in part a product of subsidized agricultural prices or incomes. This involves some fairly technical arguments, but in effect, the greatest long-term gain by farm owners from subsidies is the gain in land prices, particularly when a base acreage is tied to the land. Thus, the problems of equity, and of constraints on entry into farming, still involve the same issues raised by nineteenth-century agrarians.

Alternative or Sustainable Agriculture

Today, the advocates of alternative systems of agriculture rarely challenge fee simple ownership or advocate more compact villages, although some want to retain, or return to, smaller farms. Until the post–World War II revolution in American agriculture, the need for major alternatives scarcely existed, although the roots of what we now call organic farming predate the war. Most of those who support alternatives to what they pejoratively call industrial agriculture celebrate earlier American farms that were operated with skill and devotion. In some cases, these celebrations are colored by nostalgia or a misplaced idealism. At the same time, descriptions of contemporary, mainstream farming too often descend to caricature. Despite well-targeted criticism, the picture that emerges does not fit the self-image of most of today's successful farmers. Likewise, the almost ritualistic condemnation of official agricultural policy conceals the increasing interest in, and support for, alternative agriculture in the Department of Agriculture and among some congressional policy makers. Finally, few advocates of sustainable or even organic agriculture offer compelling explanations of how more labor-intensive and environmentally friendly approaches to farming will be able to maintain the present level of agricultural productivity.

When one looks at the history of twentieth-century agriculture, a key question is: What has changed? What has been lost and what gained? If one goes back to 1930, certain patterns are obvious. Then, most farms were mixed, in the sense that typical farmers grew several different crops and kept a variety of livestock, even when one crop was most crucial to the overall operation. Such variety was not necessarily a matter of choice but one of necessity. Until the revolutionary addition of nitrogen to fertilizer, crop rotation was a necessity for maintaining soil fertility. A rotation from corn to small grains to hay crops, preferably legumes, was the only way to avoid severe soil depletion. Although the tractor was slowly replacing horses, as late as 1930 almost all farms still used horses, at least for cultivating crops. And until herbicides were available after 1945, all row crops required cultivation. A small number of often beloved cows, hogs, and chickens were the least expensive way to provide for home consumption of meat, eggs, and dairy products. Vegetable gardens and orchards were also a near necessity.

In 1930 no one had ever dreamed of huge factory-like barns full of caged chickens or hogs. No one had heard of organic insecticides and fungicides, let alone animal antibiotics and hormones or genetically modified organisms. No one had envisioned gangs of migratory workers hired to do much of the harvesting. Such traditional farms were not organic farms in any sense of the term, but they were close to being sustainable farms, in the sense that such farming operations, if carried out with due concern for erosion control (this was a big if), could continue indefinitely, with only a few externalities to plague the larger public. But this sustainability came at a high cost—inefficiency, demanding labor inputs, low production, and low income (farmers had few off-farm employment opportunities, and farm incomes were only about half those in other sectors). For consumers, this meant proportionally much higher food costs than today. It is not surprising that, so far, financial support for alternative forms of agriculture has come primarily from affluent consumers, those who can afford to pay more for what they perceive (sometimes correctly) as safer and more flavorful food. They buy from farmers' markets, from community-based suppliers, or from Whole Foods or Wild Oats.

Almost all the organized advocates of sustainable farming want to go back to an older or, in their terms, integrated or mixed farm pattern. What they bring to this model is a much greater concern for certain environmental issues that were largely ignored or completely unrecognized by farmers in 1930, who were often indiscriminately draining wetlands or killing varmints. The words *ecology* and *biodiversity* were not part of anyone's vocabulary, and farmers were too busy, and too preoccupied with floods or droughts, to notice the beauty that surrounded them. They battled nature rather than appreciated it, but as a matter of self-respect, they did try to keep their fencerows clear, birds be damned. I became a dedicated bird-watcher in college, and some farmers probably thought I was crazy.

Today, there are dozens of organizations, many state-based, that promote what is often called "sustainable" agriculture. The word is now so popular, so widely embraced, that it always begs contextual definition. I doubt that, in the strictest sense, any system of farming is fully sustainable, because of at least a few nonrenewable inputs such as fossil fuels. But all the groups discussed in the next few pages try to minimize external inputs. They want to raise crops and livestock either without synthetic fertilizers and pesticides or with as few as possible. The goal is to retain or increase soil nutrients and always to improve the physical quality of the soil so that it will be available for future generations. At the same time, they want to avoid any environmental costs that will penalize the larger public, such as air and water pollution, damage to wildlife, soil erosion, and wetland removal. An increasing proportion of such farmers now qualify as certified organic producers, and it is the organic wing of the larger movement that now receives the most publicity. Although organic agriculture is only a subclass, not the whole, all advocates of sustainability sympathize with organic farmers and adopt many policies that overlap with organic farming. Note that almost all advocates of alternative farming are very much indebted to the research and development carried out in the past by governmental agencies. They also use the best machinery and crop varieties. In fact, alternative farming requires more technical knowledge and greater managerial skills than does conventional agriculture.

The sources of sustainable or organic agriculture are many, including Henry Carey. But the modern crusade can, arguably, be traced back only to 1924 and the intellectually versatile Rudolf Steiner, who, in a lecture the year before his death, launched what he called biodynamic agriculture. Steiner was a scientist and philosopher, a Goethe scholar, and a mystic or spiritualist (he joined the theosophist movement and believed in reincarnation). Born in 1861 in what was then southern Austria, he later moved to Switzerland, where he established the headquarters of a philosophical-religious movement.

Steiner believed that soil embodied cosmic laws and planetary rhythms. Humans could renew the force and vitality of soil by using a special type of therapeutic compost consisting of animal manure and silicates, chemically enriched by mixtures of six varieties of plant material and then seeped in water. He then advocated spraying this liquid fertilizer on growing crops. To ensure harmony with lunar and planetary cycles, Steiner's disciples worked out a rather elaborate calendar to guide the timing of crop planting, or essentially a type of astrology. His system required livestock on any farm and a diverse blend of crops, all to match the unique qualities of a given farmstead. He wanted all people to understand this type of holistic agriculture and to gain the health benefits from consuming biodynamic food. In a sense, he invented a type of community agriculture. His ideas, along with his cult-like rules, led to biodynamic associations in Europe and the United States. Some survive today and make up a tiny segment of the alternative farming movement. Steiner's form of agriculture, as developed by his disciples, often proved successful, probably because his soil improvement methods happened to overlap what is now considered good organic practice.[4]

Most advocates of organic farming, and even those espousing a less rigidly circumscribed form of sustainable agriculture, celebrate Sir Albert Howard of Great Britain as their founding father. Born in 1873 on an English farm, Howard attended Cambridge University, briefly taught in an agricultural college, and in 1905 moved to India to direct several agricultural research centers. He was a soil scientist and was knighted in recognition of his specialized research. In India he developed theories about proper farming methods, which centered on a type of composting. He was not directly influenced by Steiner and did not buy into his esoteric philosophy. Back in England in 1931, Howard began publishing articles and books about his composting system, which he referred to as the Indore process (named after a state in India). Howard believed that synthetic fertilizers ultimately harmed the soil and advocated, in their place, the recycling of all vegetable and animal wastes. Since animal manure was essential to his compost methods and the creation of humus, he believed that all farms should have a mixture of livestock and crops. In words that echoed those of Henry Carey, he talked of a law of return, by which he meant that agricultural productivity and human health required that we return as much to the soil as we take away. By World War II, his publications were well known in Britain and, increasingly, in the United States. But Howard was gradually becoming more hostile toward the existing agricultural establishment in Britain and wary of the growing role of agribusiness in farming.[5]

In 1941 Howard gained his most influential disciple in Jerome I. Rodale, the leading name in American organic farming. Rodale did not grow up on a farm. In 1941 he lived in Emmaus, Pennsylvania (near Allentown) and owned an electrical equipment dealership. He had also established Rodale Press, which published a health magazine. After reading one of Howard's books, Rodale was quickly converted to his theories and began a correspondence with him. Before the year was out, Rodale had bought a sixty-acre farm in order to try out Howard's methods. In 1942 he began publishing a new magazine called *Organic Farming and Gardening* and made Howard the associate editor. Almost no farmers subscribed, but an increasing number of backyard gardeners did. He changed the name to *Organic Gardening and Farming* and, eventually, just *Organic Gardening*. By 1980, it had 1.3 million subscribers. In 1945 Rodale published a book to promote organic farming, *Pay Dirt*, which included an introduction by Howard.

In these early years, Rodale was ignored by established farmers and often ridiculed by agricultural scientists. In 1951 his son Robert became the president of Rodale Press and an able promoter of organic methods. J. I. Rodale died in 1971, at age seventy-two. At that point, Robert bought a 305-acre farm at Kutztown and made the Rodale Research Center an experiment station for organic agriculture; it soon attracted a number of well-trained agricultural scientists. By 1985, the thriving Rodale enterprises had gained increased respect from the agricultural establishment, and Robert Rodale joined the Department of Agriculture in conducting workshops on organic farming. The Rodale Research Center sponsored affiliates abroad, and on a trip to the Soviet Union in 1990, Robert was killed in a car accident. His wife and son took over leadership of the press and the center. By then, they had become well-known American institutions with increasing acceptance and support from the government.[6]

Other early but quite heterogeneous advocates of a sustainable agriculture often sympathized with the organic movement but did not choose to adopt all its strict rules. Collectively, these critics of contemporary agriculture made up a self-conscious network, for they knew one another, gathered at meetings, and corresponded extensively. What all shared was a profound distrust of the existing forms of commercial agriculture in America. I have space for only a brief introduction of a few of these pioneers. Some pushed single ideas. One of these was Edward H. Faulkner, who published a small, best-selling book in 1943 with the arresting title *Plowman's Folly*. He painted a dismal picture of American agriculture at the end of the Great Depression, or of rampant erosion, sour soils, mounting floods from runoff, a lowering water table, even vanishing wildlife. In a sense, he was an organic farmer, for he decried the progressive loss of organic matter in American soils and the substitution of synthetic fertilizers to replace the nutrients. He blamed almost all these problems on the use of the moldboard or turning plow. It buried litter, manure, or green crops at too great a depth to allow new plantings to gain from the nutrients before they were leached deeper into the ground. Also, the smooth and clear soil on top was highly erodible. He believed most crops produced better with no plowing than with such deep plowing and urged farmers to use a shallow tillage with either disk plows or chisel plows, both of which mixed the organic material in the top three inches of soil. His elaborate, quasi-scientific defense of this theory did not persuade all readers and clearly fit some soils better than others. But it did contribute to a surge of new tilling techniques, including no-till methods that became feasible after the introduction of herbicides. Minimal tillage would become an article of faith by most advocates of a sustainable agriculture.[7]

Equally influential was novelist Louis Bromfield. Much influenced by Howard, he wanted to prove that new farming methods could restore degraded farms in his boyhood community in Ohio. With his wealth from publishing, he built a large mansion and began to restore to its original fertility the thousand-acre Malibar Farm. He spent unlimited amounts of money and was guided by both the organic methods of Howard and the most up-to-date scientific knowledge borrowed from experiment stations. He used some synthetic fertilizers along with all the compost he could gather, used different cover crops to control erosion, and reported impressive yields after a few years of rehabilitation. Yet, in the end, the farm was not profitable, and it was clear that anyone without Bromfield's wealth would be unable to duplicate his results. He admitted his failure and shifted to a more conventional form of farming. His prominence nonetheless helped advertise what he often described as an ecological form of agriculture.[8]

Other pioneers in alternative farming methods included Richard Thompson, a graduate of Iowa State. Thompson used a mixed, integrated farming system that involved most of the practices of the emerging organic movement but with one distinctive contribution—a ridge-till method. He planted crops on ridges, a system that required only minimal tilling and was suited to the colder soils of Iowa, where no-till retained so much moisture as to retard spring planting.

Wes Jackson, a persuasive writer, biologist, and geneticist, became disillusioned with conventional agriculture and established an experimental Land Institute in Salina, Kansas. He came to believe that the lush, sustainable growth of long-stem grasses on the prairie offered a possible model for a chemical-free agriculture. Jackson tried to find or breed a group of seed-producing perennial plants that could serve most of the food needs of humans, a system he labeled perennial polyculture.

In New England, John Todd helped found the New Alchemy Institute, which sought technologies to enable the type of small, intensive, village-based farming earlier praised by Henry Carey. The institute tried out different (and soon fashionable) crops, such as Jerusalem artichokes and amaranth; used greenhouses and geodesic domes; and referred to its system as permaculture.[9]

In the late 1960s and early1970s, many rebellious youth fled to the woods and countryside to set up communes. Along with their rejection of most aspects of the larger culture, they tried various new farming methods, flirted with vegetarianism, and, above all, rebelled against the unhealthy foods produced by the American agricultural establishment. Probably the most influential examples for such modern homesteaders were Scott and Helen Nearing. As a young man in 1905, Scott Nearing settled in the single-tax town of Arden, Delaware, and planted what would be classified an organic garden by later criteria (many similar gardens existed at the time). A youthful advocate of a social gospel, Nearing eventually left the church, joined the Socialist Party, condemned U.S. involvement in World War I, briefly joined the Communist Party, and even ran for mayor of New York. In 1932, already committed to the simple life, he and his wife, Helen, decided to move to a homestead in Vermont and later to one on the coast of Maine. They were dedicated vegetarians, had a deeply religious commitment to a preserved natural order, and devoted themselves to a simple life based largely on what they made and grew on their homesteads. In 1954, soon after moving to Maine, they published a book, Living the Good Life, which became very influential after they published a new edition in 1970. It attracted thousands of visitors to their homestead, who came to gain inspiration and learn how to live on the land.[10]

The most far-reaching critique of contemporary American agriculture is by Wendell Berry, a farmer who lives in Henry County in northern Kentucky. He is also a poet and a brilliant essayist and for many years was a university professor. His views are fully reactionary and, because of that, as radical in their implications as the views of any critic of our present farming system. His departure point is his direct knowledge of a relatively small, diversified tobacco farm, or the kind of farm that has almost disappeared even in his part of Kentucky. My description of my own family farm in the 1930s (in chapter 2) stands as a reference point for understanding Berry. He knows, from firsthand experience, the dark side of farming, but he chooses to idealize it at its rare best. But Berry, unlike other defenders of beleaguered small farmers, offers a critique of American culture as a whole, from which he is completely alienated. Today's maladies are not unique to contemporary agriculture but are even more deeply rooted in other sectors of the economy. The total culture is ill.

Berry idealizes a nation made up of independent, freestanding citizens. In each case, whether among farmers or craftsmen, their independence reflects their ownership of land and tools, their pride in good workmanship, and their goal of providing for families and nurturing supportive communities. For farmers, this means a love of the land and a concern for the health of those who depend on it. Modern, industrialized production, whether in a factory or on a farm, negates these simple values. Profits, growth, progress, efficiency, productivity, specialization, markets, and greed have replaced the traditional values and led to the reckless exploitation of our natural resources. Fossil energy has replaced artistry, technology has supplanted morality. Behind the changes in agriculture are those people who embrace these new values, such as the bureaucrats in the Department of Agriculture, the overly specialized scientists in our agricultural colleges, and the few highly competitive farmers destined to survive on this treadmill of progress.

Berry has embraced the idea of ecological balance, admires the work of Albert Howard, and recognizes the benefits of an organic approach, but he has a more eclectic approach to agricultural policy and fears that commercially oriented farmers are already co-opting organic techniques. He wants to support traditional farmers, with their own special knowledge about local soils and climate, such as his friends and neighbors in Kentucky. He deplores agricultural policies that have abetted only the large commercial farmers who have pushed small farmers off their land. He seriously proposed, in 1977, that small farmers go back to horses and reject tractors. He would like to retrieve local markets for local produce such as eggs, butter, and cheese. He wants farms to grow as much of their own food as possible. He knows that farmers have to make a living, but clearly he does not believe that people have half as many needs as our consumer economy pushes on them. He resents governmental regulations as much as centralized corporate farms.[11]

Berry's ideals are closest to those of a range of relatively small farmers who do not necessarily subscribe to the orthodoxy of organic farming but still try to prosper by adopting a sustainable form of farming. Some have organized, and others are simply older farmers trying to hold on to old ways. Berry knows, and idealizes, a few of those. The organizations that support such local farmers are numerous, but they have much in common. Among such groups are Sustainable Farming Associations in Minnesota and Pennsylvania, the Ecological Farming Association in California, the Practical Farmers of Iowa, the Regenerative Agriculture Association, the Institute for Alternative Agriculture, and the National Family

Farm Coalition. Animal rights organizations and several religious organizations have identified with sustainability efforts, including the Mennonite Institute for Sustainable Agriculture and the Catholic Rural Life Conference. Such organized groups are most active in the upper Midwest (particularly in Iowa, Minnesota, and Wisconsin), in New England, and on the West Coast; they are least active in the South. The farmers who formed these associations, or those who join them, have farm operations that may not be distinguishable from some traditional farms, such as small dairy farms (100 to 300 acres) that have survived in the upper Midwest or New England.[12]

What the advocates of sustainability want to regain is the rotation of both crops and pastures (with cows alternating between pastures, allowing time for recovery in the idle fields). Such rotation patterns preclude huge fields of corn or wheat and do not require the huge tractors and combines that go with them. These advocates want to protect wetlands and native species, and they often celebrate an almost mystic or religious identification with the land itself. They seek natural forms of fertilizer, use biological insecticides when possible, recycle all manure and crop residues, adopt various methods to prevent water runoff, and practice minimal tillage. Most farmers in these organizations are not organic farmers, for they may use synthetic fertilizers, herbicides, and insecticides (but as little as possible). They use the latest tools, although in a size that fits their operations. They accept whatever support they can get from federal programs. They may admire Amish or Mennonite farmers but do not join them in rejecting modern technology. They also admire certified organic farmers but are not willing to accept the stringent requirements for such certification.

One question remains. Can they make a living with alternative forms of agriculture? This depends on many factors, including location, soil type, crop selection, and the size and scale of operations. In some cases, profitability depends on market strategies. For example, one cannot meet the cost of growing free-range chickens or hogs at existing market prices. For most crops, certified organic farmers have to sell their produce above market prices to compete with conventional farmers. Effective demand for such higher-priced produce requires broad community support, as in the case of consumers willing to buy weekly baskets of food from local farmers (community-based agriculture) or patronize well-developed farmers' markets. Perhaps surprisingly, the Department of Agriculture offers the most detailed information on developing such local markets, and recent farm bills have offered grants and subsidies to help increase such markets.[13]

FEDERAL SUPPORT OF SUSTAINABLE AGRICULTURE

The official response to the movement for sustainable and organic agriculture began in the early 1980s. Because of many pressures, particularly from organized environmental organizations, both Congress and the Department of Agriculture had to respond to a growing number of critical voices. In a sense, this began in 1962 with Rachel Carson's *Silent Spring*, which placed pesticide-using farmers on the defensive. A decade later, the Department of Agriculture had to respond to another best-selling book, Jim Hightower's *Hard Tomatoes, Hard Times*. Based on a research project headed by Hightower, a former agricultural commissioner in Texas, this book contained a broadsided critique of what Hightower called the land-grant college complex, with the Extension Service a primary target. The agricultural establishment, with its close ties to agribusiness, had pushed farm laborers to a servile status, had gradually impoverished the 87 percent of small farmers, and, despite low costs, had burdened consumers with poor-tasting foods (thus the hard tomatoes of the title) and health risks from pesticides and hormones. In Hightower's words, "Genetically redesigned, mechanically planted, thinned and weeded, chemically readied and mechanically harvested and sorted, food products move out of the field and into the processing and marketing stages—untouched by human hands."[14]

In 1980 Robert Bergland, secretary of agriculture in the Carter administration, believed that the department should investigate the alternative and organic farming movements. He solicited a report, which was sympathetic. In 1981 the American Society of Agronomy concluded that organic methods could contribute to a more sustainable form of agriculture. Just before leaving office, Bergland created a new position, the organic resources coordinator. Under the new Reagan administration, an unsympathetic Secretary of Agriculture Earl Butz eliminated the position and publicly ridiculed the organic "cult." The fired coordinator joined the Henry A. Wallace Institute for Alternative Agriculture. Despite this put-down by Butz, some extension stations were doing research on organic farming. In 1984 the Michigan State Extension Service and Experiment Station joined with Rodale Press in sponsoring a conference on sustainable agriculture and integrated farming systems. In 1988 the Department of Agriculture established the small Low-Input Sustainable Agriculture Program, which used funds to support research on low-input farming. The title reflected a desire to broaden such research beyond organic farming. Its head was the same official that Butz had dismissed in 1981.[15]

The broader scientific community was increasingly aware of alternatives in agriculture. In 1989 the Board on Agriculture, a component of the National Research Council, published a report entitled *Alternative Agriculture*, which challenged conventional wisdom and scientific dogma in the Department of Agriculture. It provided the most ambitious evaluation of alternative agriculture ever completed in the United States. Perhaps not surprisingly, it was able to demonstrate, quite clearly, the environmental benefits of what could best be described as low-input agriculture. Such methods lowered the amount of nonpoint pollution, lessened soil erosion, eliminated most residues from pesticides, and improved the quality of soils. Thus, these methods could preserve the resource base, reduce costs, and protect human health. The report's list of alternative methods included crop rotation, the recycling of nutrients, integrated methods of controlling pests, and no-till or low-till farming. It found that the ablest alternative farmers produced as much per acre as conventional farmers, earned as much income, and secured major benefits for the larger society (fewer externalities). The board used eleven case studies to illustrate the diversity of such farms—from a Colorado ranch with more than 200,000 acres to a 160-acre farm in Iowa, from crop farming in the humid East to irrigated vegetable and fruit farms in the West, from farms that minimized the use of synthetic fertilizers and chemical pesticides to fully organic growers. Two of the case studies involved Richard Thompson's farm in Iowa and part of the Rodale lands in Pennsylvania.

The Board on Agriculture was very critical of existing commodity programs in the United States. In this critique, it largely agreed with the most fervent advocates of alternative farming methods. Existing policies almost forced farmers to grow the same crops year after year in order to retain their base acreage, channeled their production into subsidized crops and away from more healthy food crops without subsidies, and often encouraged farmers to expand into highly erodible land. Such pressures rewarded high production and expansion of existing methods and offered disincentives to soil-improving, disease-preventing crop rotation or mixed patterns of crops and livestock. The public not only had to pay the cost of subsidies but also had to suffer the long-term environmental consequences of such policies.[16]

Such criticisms had an impact on the Department of Agriculture and on Congress. In 1990 the department's Office of Science and Education sponsored a workshop titled "Sustainable Agriculture Research and Education in the Field," with Robert Rodale as one of the participants. In the 1990 farm bill, Congress changed the name of the tiny Low-Input

Sustainable Agriculture Program to the Sustainable Agriculture Research and Education Program (SARE), which still exists. With its limited funding, it makes research grants to a diverse group of educators, agricultural scientists, and extension agents. It publishes the research results in what it calls the *Sustainable Agriculture Network*. In 1985 the librarian of the National Agriculture Library at Beltsville, Maryland, used SARE funds to launch the Alternative Farming Systems Information Center. It is now the best source of information on alternative methods of farming, including detailed guidelines for organic farming. Its intended audience is broader than the organized sustainable or organic groups, but its bias is clearly in favor of alternative approaches. At the same time, by its original intent, the Environmental Quality Incentive Program offers grants that are congruent with the concerns and goals of alternative agriculture and is a favorite resource among many alternative farmers. Thus, it is unfair to argue that the huge bureaucracy in the Department of Agriculture favors only large commercial farmers. So far, however, funds for alternative approaches are small, though proportionate to the 1 or 2 percent they contribute toward total farm production.

CERTIFIED ORGANIC FARMING

Today, most people identify alternative farming with organic farming, and with good reason. Until 1990, organic growers usually belonged to local associations, some at the state level. These groups developed and published standards for organic farming, as did the state of California. But the label "organic" did not always have the same meaning, and there was nothing to prevent growers from using the label, even if doing so violated conventional rules. Committed organic farmers wanted some form of third-party control and successfully pressured Congress to pass the Federal Organic Foods Production Act of 1990. This act required the Department of Agriculture to certify organic producers as soon as a panel was able to agree on standards. Over the next two years the European Union enacted organic standards, but it took much longer in the United States. The American panel took a decade to complete its work because of contending factions. The first version of its standards allowed the use of sewage sludge, food irradiation, and genetically modified crops. Because of intense opposition from many organic farmers, it finally eliminated these options in the standards that took effect in 2001. It retained a few contested provisions, including permission to use synthetic micronutrients in the case of a demonstrated need (these are trace elements that may be lacking in some soils and are almost impossible to remedy with organic fertilizers).[17]

The standards are tough. Beginning in 2002, all growers who sold more than $5,000 of purportedly organic crops had to obtain official certification from the Department of Agriculture, which only then would allow them to display an organic seal. Those selling less than $5,000 could seek certification but often chose not to complete the costly certification process. In any case, these smaller growers had to follow the same standards, but without certification they could not display the seal or provide inputs to larger, certified organic growers. Some small organic farmers have protested this policy and fear that certification will lead to, in effect, a large commercial involvement in organic agriculture. The Department of Agriculture does not do the actual certifying of growers but turns this over to local, private certifying agencies, which it accredits. The farmer has to pay for the certification process, which can cost up to $2,000 and involves not only long forms but also an on-site inspection. The 2002 farm bill contained a small sum ($5 million) to provide matching grants to help farmers pay for certification.

The rules for organic growers are quite complicated and not easy to fulfill. These rules apply to at least five areas: soil maintenance, crop inputs, farming methods, care of livestock, and food processing. Farmers must commit to conserving and improving the soil. In fact, soil conditions are the key to the organic promise of healthier food. Totally apart from the organic rules, everyone who has farmed or gardened realizes the importance of the physical quality of the soil. The correct pH (acidity) and vital nutrients are only part of the equation. By adding lime (produced by the kiln firing of limestone) to sweeten the soil or sulfur compounds to raise its acidity, anyone can obtain the optimal pH (usually close to a neutral 7 on a scale from 0 to 14). Synthetic fertilizers in the right concentration can provide optimal levels of the three major nutrients. But beyond this, the soil needs abundant organic matter (provided by decaying plants, manure, or compost) to make it porous and pliable, to stimulate microbial and worm activity, to allow the mixing of water and oxygen, and to ensure a gradual release of nitrogen and other nutrients. When these conditions are all present, plants thrive. They grow rapidly, with full foliage, and gain a high degree of insect and disease resistance. They also taste good.

It is difficult to disagree with this central and obvious claim of organic farmers—organic matter is critical to soils, and the maintenance of rich, organic soils is a prescription for success in growing the most flavorful and healthy foods. What is more easily challenged is the claim that inputs of lime and synthetic fertilizers are in all cases harmful to the soil and to its products. Organic growers can, and at times must, use ground limestone to sweeten soil (unlike lime, which is a processed product, limestone is a natural ingredient that works much more slowly). They also use rock phosphate, which very slowly releases phosphates that plants can absorb. Organic farmers use several forms of organic fertilizer, including manure, compost, and ground plant or animal materials (bone, cottonseed, or alfalfa meal). The nutrients from such "natural" sources are identical to the nutrients in synthetic fertilizers, although some of the organic sources have the advantage of a generally slower release and thus less leaching into groundwater. What outsiders cannot comprehend is why a farmer, who has soils with ample organic content, cannot use faster-acting lime or synthetic fertilizer to get the needed pH or nutrients.

Organic rules prohibit any synthetic pesticides. Only natural elements such as sulfur or zinc, or biological agents such as Bt, are allowed to control insects and diseases. The rules do not allow any genetically modified crops. Generally, for vegetables, these rules do not pose a major problem. Most healthy, fast-growing vegetable crops are very resistant to both diseases and insects, but not in all cases or at all times. The problem is greatest for orchard crops. The introduction of new disease- and insect-resistant varieties of apples has helped. But in some cases, organic fruit simply is not as blemish or insect free as those treated with synthetic pesticides. In 2007 organic farmers were most concerned about a threat to one of their favorite food crops, soybeans. Asian soybean rust can destroy a crop, with the infection easily blowing from field to field. Researchers are searching for resistant varieties, but these may well involve genetic engineering and thus be unusable by organic farmers.

Weed control is a major problem in organic farming. Since herbicides are banned, some form of tilling or cultivation is often needed. Because deep cultivation invites erosion, organic farmers use shallow tilling methods or, when possible, deep mulching. The use of fire (torches) to kill weeds is allowed in the few cases where this is possible. Crop rotation can lower the amount of weeds and reduce soil diseases. Ridge-till methods may allow deeper stubble to smother weeds or make cultivation easier.

To convert to the more labor-intensive, but at times more profitable, organic mode, existing farmers have to go through a long, tough process. Since organic foods often sell for double the price of nonorganic foods, some farmers have a strong economic motive to make the switch.

For those converting to organic farming, it takes three years to clear the existing land from the effects of synthetic fertilizers or pesticides, and sometimes five years or more before yields begin to match those on conventional farms. In their application for organic certification, farmers have to agree to use specific practices that will improve the physical, chemical, and biological conditions of the soil, such as the use of cover crops, manure or compost, and crop rotation. They cannot use raw manure in the 120 days before the harvest of food crops that have direct contact with the soil, or within 90 days for other foods. Even the common practice of composting manure, despite the heating involved, does not always kill E. coli or Salmonella bacteria. Thus, organic farmers usually limit the use of unsterilized manure to cover crops grown in the year before conversion to food crops.

For organic livestock, all feed has to be organic. In the midst of the present ethanol boom, organic grain has become both scarce and very expensive. Organic farmers can use no chemical additives in feed and no antibiotics, steroids, or hormones. The rules proscribe any type of cruel treatment of animals. One has to take into account animals' natural behavior, which means that cows and sheep must have access to pasture and shade, and hogs and chickens must have room to roam in either barns or feedlots.[18]

Contamination from the outside is often a problem for organic farmers. Nearby farmers may use pesticides or graze their cattle along a boundary fence. Isolation from neighboring nonorganic farms is the best strategy, but one that is not always feasible. The possibility of contamination haunts the processing, transportation, and marketing of organic foods. The only solution is tight packing at the point of origin, clear labeling, and strong mandates for retailers to protect the integrity of the food and to display it in segregated areas. Perfect control over this process is probably impossible, for it involves too many workers who may not recognize, or care about, the problem of contamination. Thus, the surest way to maintain purity is for organic growers to market the food locally, to restaurants or to organic food markets. But this is not feasible for large growers, who often ship organic products from California to the East Coast. At present, it seems unlikely that organic foods will become an important component of food shipments to foreign countries, because of both its high price and problems of integrity.

Although growing rapidly, organic farming remains a small component of American agriculture (about 0.05 percent). In 2005, 8,493 operators grew crops on just over 4 million acres, more than double the amount in 2002. In a few vegetable crops, such as carrots, the proportion of organic produce is relatively high (7 percent), but it remains very low in grain crops and is only 1 percent in dairy farming. What is not clear is the average yield, or how organic prices compare with those gained by conventional farmers. The proposed 2007 farm bill contained funds for such data collection and analysis. In the farm census of 2002, just as the certification began, approximately 80 percent of organic farms were small, with annual sales under $50,000. Yet the 1,220 organic farms with sales over $50,000 accounted for 80 percent of the total product.

At present, I see no possibility that organic farming will become anything more than a small but rapidly growing niche in American agriculture. The rigid requirements will limit its growth, and the prohibition on synthetic fertilizers and herbicides will make it almost impossible for farmers in certain crops to adopt its methods. At the same time, the inability to use herbicides will conflict with conservation policies, particularly the increasing support for no-till cropping in many major watersheds. But in some smaller crops, particularly vegetable and fruit crops, organic farming may capture a large share of the market. Ironically, it will do so not because of the idealistic, environmentally driven ideology of the early movement but because of purely economic reasons—it can be profitable, so long as its products garner premium prices. I sense a major gap developing between the small, low-income, often part-time idealistic organic farmers who produce less than 20 percent of organic foods and the few large-scale organic farmers who produce all the rest.

I anticipate a much larger role for nonorganic, low-input, sustainable farming. In fact, the boundaries between conventional and alternative agriculture are blurring. Several current realities that are sure to become more important in the near future will make it profitable for more farmers to rotate crops, seek internal methods of preserving fertility (more recycling and more cover crops), and use integrative pest management tactics to reduce pesticide use. These realities include the end of cheap oil, and thus higher fuel and fertilizer costs; more rigorous conservation and pesticide legislation; and a warming climate, with more extreme weather events and more constraints on water use. In the twentieth century, a plentitude of resources, including good land (we still have this), abundant and cheap energy, easily mined phosphates and potash, subsidized and inexpensive water for irrigation, and an array of new and relatively inexpensive but environmentally risky pesticides, made possible the productive revolution. Except for land, all these inputs are going to be less plentiful and more costly in the twenty-first century, which should favor some of the more traditional methods pushed by alternative farming organizations. But such methods must be efficient, which in most cases will require that farming remain mechanized, scientifically informed, and chemically supported. Only such an agriculture will be able to feed 9 billion people by 2050.

Afterword

The growth in population and consumption in the twentieth century was not only unprecedented but also unrepeatable. The world population quadrupled, from 1.5 billion to more than 6 billion. The U.S. population quadrupled between 1900 and 2008, from 76 million to 304 million. The level of consumption in industrialized countries grew much faster, reaching a level that would astonish and possibly dismay earlier generations, for at least half the consumption in wealthy countries involved luxuries and all manner of frivolous goods and services. Such a continued growth rate in the twenty-first century would mean a world population of 24 billion by 2100 and an American population of around 1.1 billion. The earth's resources cannot support 24 billion people. It is almost inconceivable that the United States could support a billion people with much more than subsistence incomes and after an enormous drawdown of such finite resources as soil, water, forests, and fossil fuels.

Agriculture is the only economic sector that is tightly correlated with population. Food supplies set a limit to how many people can live on this earth. At the same time, the demand for food is so inelastic that farmers cannot profitably grow more food than people need or want to consume. Today, more than 800 million people on earth are hungry, and perhaps just as many more malnourished. In this sense, people in some regions have probed the limits of growth, and millions of people, particularly infants, have died for lack of adequate food. The Malthusian equation fits. This does not mean that the world as a whole cannot produce enough food for the present 6.5 billion people. In fact, it can, because some countries, particularly the United States, can easily expand food production. The problem is getting the food to where it is needed most.

These realities suggest a different perspective on American agriculture than I have emphasized in this book. Over and over farmers, as well as American consumers and taxpayers, have struggled to find ways to cope with too much food. Our farm problem, as we usually conceive it, has been excess production. But from a larger, and particularly a global, perspective, how lucky we have been. Even as our population has quadrupled, the total output of agricultural goods has slightly more than quadrupled, in part because we have exported anywhere from one-fourth to one-third of that product (about 28 percent in 2007). Some Americans still suffer hunger, but not because of any scarcity of food in the economy. The average American family tosses enough food in the garbage every day to feed a villager in India.

In view of such good fortune, some questions that still challenge both historians and economists may seem relatively insignificant. One is the role of American farm policy in supporting or possibly retarding the productivity revolution on American farms. Citizens, as consumers or taxpayers, have paid for various subsidies. If one adds the cost of these commodity programs to what we all pay for food, the question is whether the public has benefited from these programs or suffered a net loss in income relative to what they would have gained from a free and open market. This is only the first in a series of "what if" questions. Would farmers, in a freer market, have produced more or less food? Has the structure of production controls and price supports forced American farmers to expand production, enlarge their base acres, seek the greatest possible efficiencies, use the best and latest tools, and adopt the most potent chemicals? Have these controls and subsidies provided the type of long-term security that encouraged farmers to use credit to upgrade technologies at a much faster pace than they would have in an unregulated and often hazardous marketplace? From a different perspective, given the social costs of the mass migration of displaced farmers to cities, were the subsidies misdirected? Is it possible that income support programs for small farmers, allowing them to remain on the land and rural communities to thrive, would have contributed more to national welfare than the policies actually chosen, despite higher food costs?

I have not answered any of these questions in this book. The reason is simple: the complexities of the issues, the impossibility of engaging all the counterfactual possibilities, the inability to test any position taken, the frequent merging of issues of value and economic returns, and the lack of sufficient data to gain a high probability ranking of causal factors all intimidate me. I also lack the skills to carry out research on many of these highly technical problems. But I have gained one strong impression from my work on this book—that free-market theory, a model and not an empirical body of knowledge, does not always help one understand American agriculture. Most economists assume that governmental management of economic activity in the form of tight regulation is, in the long term, conducive to inefficiency. At the same time, they assume that special aid, or subsidies, for any class of producers will harm the overall economy. In the case of American agriculture, such model-based generalizations do not seem to fit, given that over the last seventy years agriculture has been our most efficient and most productive sector. It is possible that it would have been even more productive without governmental guidance and help, but before buying into that proposition, I would like to see proof.

In this book, I have pointed out how our farmers more than met the challenge of a rapidly growing population. It is remarkable that they could increase food production fourfold, and on roughly the same number of cultivated acres, even as the labor expended slowly declined to one-fourth the amount required in 1900. It is equally remarkable that the oldest tool ever used by humans, fire, made this possible. The controlled combustion of hydrocarbons that cooked food and warmed caves in the distant past now powers combines and extracts nitrates from the air. Decaying vegetation, over millions of years, bequeathed to us a huge storehouse of sequestered hydrocarbons that, in the form of coal, natural gas, and, above all, petroleum, fueled the growth of the twentieth century. It was a century of cheap oil, as humans drew down this bank of energy by a greater amount than in all past history. The age of oil is not over, but in all probability, cheap oil is a thing of the past. Before the end of this century, the oil that is now available by conventional methods, and at a price that encourages its use, will be gone, and natural gas will be in short supply. This is only one of many challenges faced by humans, and particularly by farmers. Water shortages and possibly major shifts in climate are two others.

Yet, despite all the challenges, American farmers are better situated to meet them than are farmers anywhere else in the world. We still have plenty of good soil. Our population may rise to 400 million by 2050, before leveling off. Our present birth rate is at a replacement level, or higher than in any other industrialized country. The projected growth largely involves immigration. Such growth may lead to multiple problems, including major environmental ones, but the food supply is not at risk. Our farmers, even at present, could feed 400 million. They could do it more easily if food habits shifted away from meat and toward more vegetables. They might have to clear some forests, end conservation reserves, shift land from such nonfood crops as cotton and tobacco, stop the diversion of food crops into biofuels, and possibly even renew cultivation in highly erodible areas, but they could do the job without major shifts in the structure of farming. If Americans had to, they could plant gardens in suburban yards, convert golf courses into fields, or actively farm strips along highways.

I doubt that farmers will be able to maintain the pace of growth that has blessed consumers over the last half century. As always, farmers will continue to produce enough food to keep up with population growth, but probably with fewer surpluses. The level of growth in bushels or pounds of food will be largely determined by population growth in the United States, with one qualification—it could include a much larger volume of exports to help feed the less fortunate parts of the earth. The word growth is always a bit loaded. For agriculture, it can refer to either the volume of products or their monetary value. If farmers become even more efficient, they might well meet the food needs of a growing population, yet at prices that would lower the "value" of what they produce relative to other goods and services. If they can do this without increased externalities, such as environmental damage, then everyone would benefit, assuming an equitable distribution of the income received. In part, it was just such a process that so benefited consumers, and at least highly competitive farmers, over the last half century. Can such a pattern continue?

No one can be sure. But there are reasons to doubt it. In fact, it seems likely, from the perspective of 2008, that food prices will continue to rise, reversing a century-old pattern. This is because of much higher input costs for energy, irrigation works, fertilizers, and many chemicals. The recent upsurge in extreme weather events, including a record drought in the Southeast and Southwest in 2007, may be a sign of things to come if, as seems likely, the pace of global warming continues.

The surge in productivity that follows the introduction of new technologies, such as the combine or herbicides, eventually flattens out. Only comparable new tools can maintain the pace of growth. It is always dangerous to conclude that there will be no such comparable innovations in the near future. At present, farmers are benefiting not from dramatic new machines but from a gradual improvement in existing ones. If anything, because of new environmental constraints, the benefits of farm chemicals will stabilize or even decline. Gains from plant and animal breeding may have peaked already; they have certainly slowed. The one, possibly dramatic exception is the present growth in genetically modified crops, a growth that may flounder on consumer resistance or the discovery of some long-term genetic danger to humans. Of course, dramatic technological breakthroughs could occur at almost any time, either in agriculture or in needed inputs. For example, the early development of fusion power could lower the costs of electricity, success in achieving a cost-effective method of producing cellulosic ethanol might open up new markets for farmers, and new battery technologies could make electric tractors and combines a reality. Also possible are new managerial skills and additional gains from economies of scale. That is, farm consolidation could continue, and the number of farm operators could continue to decline.

I am not sure that higher food costs are undesirable, but I have to qualify such a judgment. In an economy as prosperous as that of the United States, it seems appropriate that food should cost more than it does at present. It would be nice if that more expensive food was healthier than indicated by present American diets. When American consumers as a whole are building houses more than double the size of 1950s homes, keeping two or three automobiles in every garage, spending more on recreation than the annual national income in more than half the world's nations, paying more than $100 million a year to compensate baseball players and billions to pay chief executives, providing dozens of toys for every child, indulging in every conceivable electronic innovation, and even throwing away $20 billion annually on bottled water, why not shift some of the national income back to food?

Here are the qualifications. First, the shift to higher costs should be based in large part on the pricing of as many externalities as possible in agriculture. Consumers of food should be willing to pay (subsidize?) farmers to reduce air and water pollution, control erosion, preserve soil quality, treat livestock humanely (no more drugged and cheap chickens), and pay higher wages and improved benefits to farm laborers. If this sounds like a prescription for the types of alternative agriculture described in chapter 8, so be it.

The other qualification involves income policy. Based on average incomes, the United States can afford a more expensive and healthier food system, but with the present income distribution, such higher costs would threaten the livelihoods of those in the lower one-third in terms of income. Thus, a more responsible, environmentally friendly agriculture would require a larger and more costly food aid system. Americans are certainly wealthy enough to afford this change, but whether they are willing to make the political decisions necessary to implement it is doubtful.

Notes

American Agriculture before 1930

1. For most estimates about agricultural growth, I relied on the excellent book by Bruce L. Gardner, *American Agriculture in the Twentieth Century: How It Flourished and What It Cost* (Cambridge, Mass.: Harvard University Press, 2002).

2. U.S. Department of Agriculture, National Agricultural Statistics Service, 2002 *Census of Agriculture*, vol. 1, ch. 1, U.S. National Data, tables 1 and 2.

3. In *Sowing Modernity: America's First Agricultural Revolution* (Ithaca, N.Y.: Cornell University Press, 1997), Peter D. McClelland both describes and illustrates the major farm tools used before the Civil War. R. Douglas Hurt provides an equally vivid record in *American Farm Tools, from Hand Power to Steam Power* (Manhattan, Kans.: Sunflower University Press, 1982). All the tools I mention, before the modern tractor, are illustrated in these two books.

4. The complex story of harvesting and threshing equipment for small grains is well told and illustrated in Graeme Quick and Wesley Buchele, *The Grain Harvesters* (St. Joseph, Mo.: American Society of Agricultural Engineers, 1978). For the special conditions in the hilly areas of Washington and Oregon and for great photographs, see Kirby Brumfield, *This Was Wheat Farming: A Pictorial History of the Farms and Farmers of the Northwest Who Grew the Nation's Bread* (Seattle: Superior Publishing Company, 1968).

5. Old tractors are now collector's items, with numerous guides to the various types and makes. For my summary I relied largely on Robert C. Williams, *Fordson, Farmall, and Poppin' Johnny: A History of the Farm Tractor and Its Impact on America* (Urbana: University of Illinois Press, 1987).

6. Gladys L. Baker, Wayne D. Rasmussen, Vivian Wiser, and Jane M. Porter, *Century of Service: The First Hundred Years of the United States Department of Agriculture* (Washington, D.C.: Government Printing Office, 1963); R. Grant Seals, "The Formation of Agricultural and Rural Development with Emphasis on African-Americans. II. The Hatch-George and Smith-Lever Acts," *Agricultural History* 65 (spring 1991): 12–34; Jane M. Porter, "Experiment Stations in the South, 1877–1940," *Agricultural History* 53 (January 1979): 84–101; Lou Ferleger, "Uplifting American Ag-

riculture: Experiment Station Scientists and the Office of Experiment Stations," *Agricultural History* 64 (spring 1990): 5–23.

7. Jeffrey W. Moss and Cynthia B. Lass, "A History of Farmers' Institutes," *Agricultural History* 62 (spring 1988): 150–63.

8. Roy V. Scott, *The Reluctant Farmer: The Rise of Agricultural Extension to 1914* (Urbana: University of Illinois Press, 1970).

9. Wayne Rasmussen, *Taking the University to the People: Seventy-five Years of Cooperative Extension* (Ames: Iowa State University Press, 1989).

10. The National Vocational Education (Smith-Hughes) Act, Public Law No. 347, Sixty-fourth Congress, S-703 (1917).

11. William D. Rowley, *The Bureau of Reclamation: Origins and Growth to 1945* (Washington, D.C.: Bureau of Reclamation, 2006).

12. The best analysis of the Federal Farm Loan Act that I have read was almost contemporaneous with the bill: C. W. Thompson, "The Federal Farm Loan Act," *The American Economic Review, Supplement, Papers and Proceedings of the Twenty-ninth Annual Meeting of the American Economics Association* 7 (March 1917): 115–31.

13. John Mark Hansen, *Gaining Access: Congress and the Farm Lobby, 1919–1981* (Chicago: University of Chicago Press, 1991).

14. Gilbert C. Fite, *George N. Peek and the Fight for Farm Parity* (Norman: University of Oklahoma Press, 1954).

15. Wayne Rasmussen, *Farmers, Cooperatives, and USDA: A History of Agricultural Cooperative Service*, Agricultural Information Bulletin 621 (Washington, D.C.: Government Printing Office, 1991).

16. Joan Hoff Wilson, "Hoover's Agricultural Policies, 1921–1928," *Agricultural History* 51 (January 1977): 335–61; David E. Hamilton, *From New Day to New Deal: American Farm Policy from Hoover to Roosevelt, 1928–1933* (Chapel Hill: University of North Carolina Press, 1991).

A New Deal for Agriculture, 1930–1938

1. The most detailed history of the Farm Board is in David E. Hamilton, *From New Day to New Deal: American Farm Policy from Hoover to Roosevelt, 1928–1933* (Chapel Hill: University of North Carolina Press, 1991). A good overview is in Gilbert C. Fite, *George N. Peek and the Fight for Farm Parity* (Norman: University of Oklahoma Press, 1954), 221–42.

2. A brief history of the effort to gain a domestic allotment plan is in Theodore Saloutos, *The American Farmer and the New Deal* (Ames: Iowa State University Press, 1982), 34–49. The most detailed and thoughtful account is in Richard S. Kirkendall, *Social Scientists and Farm Politics in the Age of Roosevelt* (Columbia: University of Missouri Press, 1966), 11–60. Also see Van L. Perkins, *Crisis in Agriculture: The Agricultural Adjustment Administration and the New Deal* (Berkeley: University of California Press, 1969), 36–78.

3. Almost all books and articles on the Agricultural Adjustment Act stress the domestic allotment scheme and slight the many other strategies in this omnibus

bill. Perkins, Crisis in Agriculture, at least lists the other provisions, as does Murray R. Benedict, Farm Policies of the United States, 1790–1950 (New York: Twentieth Century Fund, 1953).

4. The most detailed account of such marketing agreements is in Perkins, Crisis in Agriculture, 148–67.

5. C. Roger Lambert, "Want and Plenty: The Federal Surplus Relief Corporation and the AAA," Agricultural History 46 (July 1972): 390–400; Benedict, Farm Policies of the United States, 380–81.

6. Bruce L. Gardner, American Agriculture in the Twentieth Century: How It Flourished and What It Cost (Cambridge, Mass.: Harvard University Press, 2002), 217–19.

7. Kirkendall, Social Scientists, 75–83.

8. Ibid., 165–217; Jess Gilbert, "Democratic Planning in Agricultural Policy: The Federal-County Land-Use Planning Program," Agricultural History 70 (spring 1996): 233–50.

9. I explain these programs in my first book, Paul K. Conkin, Tomorrow a New World: The New Deal Community Programs (Ithaca, N.Y.: Cornell University Press, 1959), 131–85.

10. Vernon W. Ruttan, "The TVA and Regional Development," in TVA: Fifty Years of Grass-Roots Bureaucracy, ed. Erwin C. Hargrove and Paul K. Conkin (Urbana: University of Illinois Press, 1983), 150–63.

11. Sandra S. Batie, "Soil Conservation in the 1980s: A Historical Perspective," Agricultural History 59 (April 1985): 107–23.

12. Benedict, Farm Policies of the United States, 350–52.

13. Ibid., 375–78, 381–83.

14. This conclusion derives from the arguments of economist D. Gale Johnson, who concludes that artificially higher prices would provide only short-term income increases for farmers, since competition in an increasingly integrated labor and capital market would soon lower net incomes to levels comparable to those in other sectors. See D. Gale Johnson, "The Performance of Past Policies: A Critique," in Alternative Agricultural and Food Policies, ed. Gordon C. Rausser and Kenneth R. Farrell (Washington, D.C.: Giannini Foundation of Agricultural Economics, University of California, 1985), 11–36.

Dimensions of an Agricultural Revolution

1. Bruce L. Gardner, American Agriculture in the Twentieth Century: How It Flourished and What It Cost (Cambridge, Mass.: Harvard University Press, 2002), 28–47; Sally H. Clark, Regulation and the Revolution in United States Farm Productivity (New York: Cambridge University Press), 3–6.

2. U.S. Department of Agriculture, National Agricultural Statistics Service, 2002 Census of Agriculture, vol. 1, ch. 1, U.S. National Level Data, table 57 (Summary by Combined Government Payments and Market Value of Agricultural Products Sold, 2002).

3. Gardner, *American Agriculture*, 15–18.

4. *2002 Census of Agriculture*, table 55 (Summary by Size of Farm: 2002).

5. For a well-illustrated history of grain harvesting machines, see Graeme Quick and Wesley Buchele, *The Grain Harvesters* (St. Joseph, Mo.: American Society of Agricultural Engineers, 1978).

6. Gilbert C. Fite, "Mechanization of Cotton Production since World War II," *Agricultural History* 54 (January 1980): 190–207.

7. *2002 Census of Agriculture*, table 53 (Women Principal Operators—Selected Farm Characteristics: 2002 and 1997) and table 55.

8. Donald Holley, *The Second Great Emancipation: The Mechanical Cotton Picker, Black Migration, and How They Shaped the Modern South* (Fayetteville: University of Arkansas Press, 2000), offers the more irenic view, while Pete Daniel presents the dark side of mechanization in *Breaking the Land: The Transformation of Cotton, Tobacco, and Rice Cultures since 1880* (Urbana: University of Illinois Press, 1985), 155–83, 239–55.

9. Alan I. Marcus, "Setting the Standard: Fertilizers, State Chemistry, and Early National Commercial Regulation, 1880–1887," *Agricultural History* 61 (winter 1987): 47–73.

10. Richard C. Sheridan, "Chemical Fertilizers in Southern Agriculture," *Agricultural History* 53 (January 1979): 308–18.

11. Vernon W. Ruttan, "The TVA and Regional Development," in *TVA: Fifty Years of Grass-Roots Bureaucracy*, ed. Erwin C. Hargrove and Paul K. Conkin (Urbana: University of Illinois Press, 1983), 150–63.

12. Gardner, *American Agriculture*, 22–25.

13. Gail E. Peterson, "The Discovery and Development of 2,4-D," *Agricultural History* 41 (July 1967): 243–54.

14. Charles E. Little, *Green Fields Forever: The Conservation Tillage Revolution in America* (Washington, D.C.: Island Press, 1987).

15. Terry G. Summons, "Animal Food Additives, 1940–1966," *Agricultural History* 42 (October 1968): 305–13.

16. If one wants to follow these controversies, the Internet is the best resource, for it contains dozens of references to this ongoing debate.

17. Dominic Hogg, *Technological Change in Agriculture: Locking in to Genetic Uniformity* (London: Macmillan Press, 2000), 145–74.

Surpluses and Payments, 1954–2008

1. The most complete history of the Brannan Plan is by Virgil W. Dean, *An Opportunity Lost: The Truman Administration and the Farm Policy Debate* (Columbia: University of Missouri Press, 2006).

2. For the Soil Bank and all the commodity programs from 1954 to 1973, the most definitive and informed account is by a participant in many of the policy debates, Willard W. Cochrane. See Willard W. Cochrane and Mary E. Ryan, *American Farm Policy, 1948–1973* (Minneapolis: University of Minnesota Press, 1976),

175–255; Gilbert Fite, *American Farmers: The New Minority* (Bloomington: Indiana University Press, 1981), 102–19.

3. Ibid.; James M. Giglio, "New Frontier Agricultural Policy: The Commodity Side, 1961–1963," *Agricultural History* 61 (summer 1987): 36–52.

4. Bruce L. Gardner, *American Agriculture in the Twentieth Century: How It Flourished and What It Cost* (Cambridge, Mass.: Harvard University Press, 2002), 84–87.

5. Willard W. Cochrane and C. Ford Runge, *Reforming Farm Policy: Toward a National Agenda* (Ames: Iowa State University, 1992), 52–53; Ronald D. Knutson, J. B. Penn, and William T. Boehm, *Agricultural and Food Policy*, 3rd ed. (Englewood Cliffs, N.J.: Prentice Hall, 1995), 256–58.

6. Knutson et al., *Agricultural and Food Policy*, 257–58.

7. Geoffrey S. Becker, "Farm Support Programs and World Trade Commitments," in *Farm Economic Issues*, ed. Denise L. Sandrino (New York: Nova Science Publishers, 2004), 55–78.

8. The farm bill of 1996 (Public Law No. 104-127) is available at www .usda.gov/farmbill1997. I did not read all of this huge bill, but only the sections related to commodity programs. The public debates about the bill led to hundreds of articles and much criticism.

9. Jasper Womach, "Loans and Loan Deficiency Payments," in Sandrino, *Farm Economic Issues*, 183–201.

10. Ralph M. Chite, "Farm Disaster Assistance," ibid., 79–86.

11. The farm bill of 2002 is available at www.usda.gov/farmbill2002.

12. Cochrane and Runge, *Reforming Farm Policy*, 81–82.

13. Knutson, *Agricultural and Food Policy*, 430–40.

Farming in the Twenty-first Century: Status and Challenges

1. U.S. Department of Agriculture, National Agricultural Statistics Service, *2002 Census of Agriculture*, vol. 1, ch. 1, U.S. National Level Data, table 1, (Historical Highlights: 2002 and Earlier Census Years).

2. Ibid., table 52 (Selected Characteristics by Race); table 61 (Summary by Tenure of Principal Operator and by Operators on Farm).

3. These conclusions reflect an analysis of data in ibid., table 57 (Summary by Combined Government Payments and Market Value of Agricultural Products).

4. Bruce L. Gardner, *American Agriculture in the Twentieth Century: How It Flourished and What It Cost* (Cambridge, Mass.: Harvard University Press, 2002), 64–66, 107–9.

5. *2002 Census of Agriculture*, table 52.

6. U.S. Department of Agriculture, National Agricultural Statistics Service, "Farm Labor," May 18, 2007.

7. Gardner, *American Agriculture*, 87–90.

8. U.S. Department of Agriculture, "2007 Farm Income Report," February 2007.

9. 2002 *Census of Agriculture*, table 7 (Income from Farm-Related Sources: 2002 and 1997).

10. Ibid., table 3 (Economic Class of Farms by Market Value of Agricultural Products Sold and Government Payments).

11. Ibid., table 5 (Net Cash Farm Income of the Operations and Operators).

12. Willard W. Cochrane, *The Curse of American Agricultural Abundance* (Lincoln: University of Nebraska Press, 2003).

13. In several books and articles, Pete Daniel has thoroughly documented the plight of African American and small-scale white farmers in the South, beginning with *The Shadow of Slavery: Peonage in the South, 1901–1969* (Urbana: University of Illinois Press, 1972) and continuing in *Breaking the Land: The Transformation of Cotton, Tobacco, and Rice Cultures since 1880* (Urbana: University of Illinois Press, 1985).

14. Sandra S. Batie, "Soil Conservation in the 1980s: A Historical Perspective," *Agricultural History* 59 (April 1985): 107–23.

15. Luther Tweeten, *Farm Policy Analysis* (Boulder, Colo.: Westview Press, 1989), 261–88.

Alternatives

1. Henry C. Carey, *The Past, the Present, and the Future* (reprint, New York: Augustus M. Kelley, 1967), 110–12, 128–33; Paul K. Conkin, *Prophets of Prosperity: America's First Political Economists* (Bloomington: Indiana University Press, 1980), 283–85, 290–93.

2. John A. Hostetler, *Hutterite Society* (Baltimore: Johns Hopkins University Press, 1974); Paul K. Conkin, *Two Paths to Utopia: The Hutterites and the Llano Colony* (Lincoln: University of Nebraska Press, 1964).

3. Conkin, *Prophets of Prosperity*, 222–59.

4. The best introduction to biodynamic agriculture that I have found is by Herbert H. Koepf, Bo D. Pettersson, and Wolfgang Schaumann, *Bio-Dynamic Agriculture: An Introduction* (Spring Valley, N.Y.: Anthroposophic Press, 1976).

5. Sir Albert Howard, *The Soil and Health: A Study of Organic Agriculture* (New York: Devin-Adair, 1947); Sir Albert Howard, "Maintaining Fertility—The Indore Process," in *The Organic Tradition: An Anthology of Writings on Organic Farming, 1900–1950*, ed. Philip Conford (Bideford, England: Green Books, 1988), 114–31.

6. As a longtime reader of *Organic Gardening*, I think I understand what the Rodales accomplished. The best scholarly defense of their organic methods is by a successor of Robert Rodale, Richard R. Harwood, "The Integrative Efficiencies of Cropping Systems," in *Sustainable Agriculture and Integrated Farming Systems*, ed. Thomas C. Edens, Cynthia Fridgen, and Susan L. Battenfield (East Lansing: Michigan State University Press, 1985), 64–75.

7. Edward H. Faulkner, *Plowman's Folly* (Norman: University of Oklahoma Press, 1943).

8. Louis Bromfield, *Out of the Earth* (New York: Harper and Brothers, 1948).

9. Randel S. Beeman and James A. Pritchard, *A Green and Permanent Land: Ecology and Agriculture in the Twentieth Century* (Lawrence: University Press of Kansas, 2001), 123–29.

10. Helen Nearing and Scott Nearing, *Living the Good Life* (Harborside, Me.: Social Science Institute, 1954); Rebecca Kneale Gould, *At Home in Nature: Modern Homesteading and Spiritual Practice in America* (Berkeley: University of California Press, 2005), xiv–xv, 139–70.

11. Wendell Berry, *The Unsettling of America: Culture and Agriculture* (San Francisco: Sierra Club Books, 1977).

12. Beeman and Pritchard, *A Green and Permanent Land*, 131–46; Gary Holthaus, *From the Farm to the Table: What All Americans Need to Know About Agriculture* (Lexington: University Press of Kentucky, 2006).

13. Holthaus, *From the Farm to the Table*.

14. Jim Hightower and Susan DeMarco, *Hard Tomatoes, Hard Times: A Report of the Agribusiness Accountability Project on the Failure of America's Land Grant College Complex* (Cambridge, Mass.: Schenkman, 1973), 30.

15. Joseph Heckman, "A History of Organic Farming—Transitions from Sir Albert Howard's War in the Soil to the USDA National Organic Program," *Renewable Agriculture and Food Systems* 21 (September 2006): 143–50 (formerly *American Journal of Alternative Agriculture*).

16. National Research Council, Board on Agriculture, *Alternative Agriculture* (Washington, D.C.: National Academy Press, 1989).

17. Heckman, "History of Organic Farming."

18. The regulations governing organic certification are available at www.ams.usda.gov/nop/NOP/standards.html. This home page will guide you to the Electronic Code of Federal Regulations, Title 7, part 205—National Organic Program. This is an enormously complex body of rules, which I have tried to summarize.

Index